Studies in the

Evolutionary Psychology of Feeling

By

HIRAM M. STANLEY

Member of the American Psychological Association

London

1895

ISBN-13: 978-1724474261

ISBN-10: 172447426X

PREFACE

This work does not profess to be a treatise on the subject of feeling, but merely a series of studies, and rather tentative ones at that. I have attempted to deduce from the standpoint of biologic evolution the origin and development of feeling, and then to consider how far introspection confirms these results. I am well aware that I traverse moot points—what points in psychology are not moot?—and I trust that the position taken will receive thorough criticism. I should be very glad to have new facts adduced, whatever way they may bear. I have no theory to defend, but the results offered are simply the best interpretation I have as yet been able to attain.

Some of the material of this book has appeared during the last ten years in the pages of *Mind*, *Monist*, *Science*, *Philosophical Review* and *Psychological Review*, but my contributions to these periodicals have in many cases been largely re-written.

Hiram M. Stanley.

Lake Forest, Illinois, U S.A.

TABLE OF CONTENTS

CHAPTER I

ON THE INTROSPECTIVE STUDY OF FEELING

Of all the sciences psychology is, perhaps, the most imperfect. If a science is a body of knowledge obtained by special research and accepted by the general *consensus* of specialists, then psychology is so defective as to scarcely merit the name of science. This want of *consensus* is everywhere apparent, and must especially impress any one who compares the lack of harmony in manuals of psychology with the practical unanimity in manuals of botany, geology, physics, and other sciences. Even in the most fundamental points there is no agreement, as will be evident in a most summary statement.

It is now something more than a century since the general division of psychic phenomena into intellect, feeling and will, first came into repute, but still some psychologists of note do not agree to this fundamental classification, but would unite feeling and will in a single order. As to the subdivisions of feeling and will we are confessedly wholly at sea. In intellect it is only on the lower side, sensation and perception, that anything of great scientific value has been accomplished; and even now it cannot be said that the classes of sensation have been marked off with perfect certainty. In the higher range of intellect psychology can do scarcely more than accept 2some ready-made divisions from common observation and logic. And if so little has been settled in the comparatively simple work of a descriptive classification of the facts of mind, we may be assured that still less has been accomplished toward a scientific *consensus* for the laws of mind. Weber's law alone seems to stand on any secure basis of experiment, but its range and meaning are still far from being determined. Even the laws of the association of ideas are still the subjects of endless controversy. Also in method there is manifestly the greatest disagreement. The physiological and introspective schools each magnify their own methods, sometimes so far as to discredit all others. Physiological method has won for itself a certain standing, indeed, but just what are its limitations is still far from being settled.

But the grievous lack of generally accepted results is most apparent in the domain of feeling. The discussion of feeling in most manuals is very meagre and unsatisfactory. Professor James's recent treatise, for instance, gives some 900 pages to the Intellect, and about 100 pages each to Feeling and Will. There is little thorough analysis and no perfected inductive classification. We often, indeed, find essays of literary value which appeal to the authority of literature. But to refer to Shakspeare or Goethe as psychological authorities, or in illustration or proof of psychological laws, is generally a doubtful procedure. The literary and artistic treatment of human nature is quite distinct from the scientific, and literature and art cannot be said to be of much more value for psychology than for physics, chemistry, or biology. To appeal to the Bible or Shakspeare in matters psychological, is usually as misleading as to consult them for light on geology or botany. Even the fuller treatises on the subject of feeling rarely reach beyond literary method and common observation, being for the most part a collection and arrangement of the results of common sense, 3accepting common definitions, terms, and classifications. Now, science is always more than common sense and common perception, it is uncommon sense; it is an insight and a prolonged special investigation which penetrates beneath the surface of things and shows them in those inner and deeper relations which are entirely hid from general observation. Common views in psychology are likely to be as untrustworthy as in physics or astronomy, or any other department. Science must, indeed, start with common sense, but it does not deserve the name of science till it gets beyond it.

Again, the subject of pleasure, pain, and emotion, is usually discussed with considerable ethical or philosophical bias. The whole subject of feeling has been so naturally associated with ethics and philosophy from the earliest period of Greek thought that a purely colourless scientific treatment is quite difficult. Furthermore, feeling has been too often discussed from an *a priori* point of view, as in the rigid following out of the Herbartian theory of feeling as connected with hindrance or furtherance of representation. Still further, the physical side of emotion has been so emphasized by the physiological school as to distract attention from purely psychological investigation.

It is obvious, then, on the most cursory review, that very little has been accomplished in the pure psychology of feeling. Here is a region almost unexplored, and which, by reason of the elusiveness and obscurity of the phenomena, has seemed to some quite unexplorable. Dr. Nahlowsky truly remarks, that feeling is a "strange mysterious world, and the entrance to it is dark as to Hades of old." Is there any way out of this darkness and confusion? If the study of feeling is to become scientific, we must, I think, assume that all feeling is a biological function governed by the general laws of life and subject in origin and development to the law of struggle for existence. Assuming this strictly scientific point of view, we 4have to point out some difficulties in the way of the introspective psychology of feeling as compared with other departments of biological science.

We trace directly and with comparative ease any physiological organ and function from its simplest to its most complex form; for example, in the circulation of the blood there is clearly observable a connected series from the most elementary to the most specialized heart as developed through the principle of serviceability. In some cases, as in the orohippus, a form in the evolution of the horse, we are able to predict an intermediate organism. Psychology is still far from this deductive stage; we have no analogous series of psychic forms, much less are able to supply, *a priori*, the gaps in a series. The reason for this is mainly the inevitable automorphism of psychological method. In biology we are not driven to understand life solely through analogy with our own life, but in psychology mind in general must be interpreted through the self-observation of the human mind. In biology we see without effort facts and forms of life most diverse from our own; the most strange and primitive types are as readily discernible as the most familiar and advanced, the most simple as the most complex. We study a fish just as readily as a human body, but the fish's mind—if it has any—seems beyond our ken, at least is not susceptible of direct study, but a matter for doubtful inference and speculation. Whether a given action does or does not indicate consciousness, and what kind of consciousness, this is most difficult to determine. Thus we have the most various interpretations, some, as Clifford, even going so far as to make psychic phenomena universal in matter, others, on the other hand, as Descartes, limiting them to man alone.

The difficulty of this subjective method, this reflex investigation, is almost insurmountable. Consciousness must act as both revealer and revealed, must be a light which enlightens itself. A fact of consciousness to be 5known must not simply exist like a physical fact or object, as a piece of stone, but it must be such that the observing consciousness realizes or re-enacts it. To know the fact we must have the fact, we must *be* what we *know*. Mind is pure activity; we do not see an organ and ask what it is for, what does it do; but we are immediately conscious of consciousness as activity, and not as an objective organ. We must here, then, reverse the general order and know the activity before we can identify the organ as a physical basis.

By the purely objective vision of the lower sciences we can easily determine a genetic series of forms most remote from our own life, but in psychology, mind can be for us only what mind is in us. The primitive types of psychosis are, no doubt, as remote and foreign from our own as is the primitive type of heart or nervous system from that of man's. In the case of heart and nerve we can objectively trace with certainty the successive steps, but in endeavouring to realize by subjective method the evolution of mind we are involved in great doubt and perplexity. How can we understand an insect's feelings? How can we appreciate minds which are without apprehension of object, though there is reason to believe such minds exist? Only to a very limited extent can a trained and sympathetic mind project itself back into some of its immediately antecedent stages. Consciousness, because of its self-directive and self-reflective power, is the most elastic of functions, yet it can never attain the power of realizing all its previous stages. Sometimes, however, the mind in perfect quiescence tends to relapse into primitive modes, which may afterward be noted by reflection, but such occasions are comparatively rare. The subjective method means a commonalty of experience which is often impossible to attain. Thus a man may believe there are feelings of maternity; he has observed the expression of nursing mothers, and knows in a general way that here is a peculiar psychosis 6into which he can never enter, and which is, therefore, beyond his scientific analysis. The psychic life of the child is more akin to his than that of the mother; yet it is only by an incessant cultivation of receptivity and repression of adult propensities that one can ever attain any true inkling of infant experience. There is then, I think, a vast range of psychic life which must for ever lie wholly hidden from us, either as infinitely below or infinitely above us; there is also an immense realm where we can only doubtfully infer the presence of some form of consciousness without being able to discriminate its quality, or in exceptional cases to know it very partially; and there is but a relatively small sphere where scientific results of any large value may be expected. By reason of its objective method the realm of physical science is practically illimitable, but psychic science is, by reason of its subjective method, kept for ever within narrow boundaries.

We must then take into account the inherent difficulties of the subjective method as applied to the study of feeling and mind in general, and yet we must recognise its necessity. No amount of objective physiological research can tell us anything about the real nature of a feeling, or can discover new feelings. Granting that neural processes are at the basis of all feelings as of all mental activities, we can infer nothing from the physiological activity as to the nature of the psychic process. It is only such feelings and elements as we have already discovered and analyzed by introspection that can be correlated with a physical process. Nor can we gain much light even if we suppose—which is granting a good deal in our present state of knowledge—that there exists a general analogy between nerve growth and activity, and mental operations. If relating, *i.e.*, cognition, is established on basis of inter-relation in brain tissue, if every mental connecting means a connecting of brain fibres, we might, indeed, determine the number of thoughts, but we could not tell what the 7thoughts were. So if mental disturbance always means bodily disturbance, we can still tell nothing more about the nature of each emotion than we knew before. We must first know fear, anger, etc., as experiences in consciousness before we can correlate them with corporeal acts.

Is now this necessarily subjective method peculiarly limited as to feeling? Can we know feeling directly as psychic act or only indirectly through accompaniments? Mr. James Ward (*vide* article on Psychology in the *Encyclopædia Britannica*, p. 49, cf. p. 71) remarks that feelings cannot be known as objects of direct reflection, we can only know *of* them by their effects on the chain of presentation. The reason for this is, that feeling is not presentation, and "what is not presented cannot be re-presented." "How can that which was not originally a cognition become such by being reproduced?"

It cannot. But do we need to identify the known with knowing, in order that it may be known? Must feeling be made into a cognition to be cognized? It is obvious enough that no feeling can be revived into a representation of itself, but no more can any cognition or any mental activity. Revival or recurrence of consciousness can never constitute consciousness of consciousness which is an order apart. If cognition is only presentation and re-presentation of objects, we can never attain any apprehension of consciousness, any cognition of a cognition or of a feeling or of a volition, for they are all equally in this sense subjective acts. Re-presentation at any degree is never by itself sense of re-presentation or knowledge of the presentation.

Of course, the doctrine of relativity applies to introspection as to all cognition, and subject *qua* subject is as unknowable as object *qua* object. We do not know feeling in itself, nor anything else in itself, the subjective like the objective *ding an sich* is beyond our ken. Yet kinds 8of consciousness are as directly apprehended and discriminated as kinds of things, but the knowing is, as such, distinct from the known even when knowing is known. Here the act knowing is not the act known and is different in value. The object known is not, at least from the purely psychological point of view, ever to be confounded with the knowing, to be incorporated into cognition by virtue of being cognized. Feeling, then, seems to be as directly known by introspection and reflection as any other process. It is not a hypothetical cause brought in by the intellect to explain certain mental phenomena, but it is as distinctly and directly apprehended as cognition or volition.

The distinction between having a feeling and knowing a feeling is a very real one, though common phraseology confuses them. We say of a brave man, he never knew fear; by which we mean he never feared, never experienced fear, and not that he was ignorant of fear. Again, in like manner, we say sometimes of a very healthy person, he never knew what pain was, meaning he never felt pain. These expressions convey a truth in that they emphasize that necessity of experience in the exercise of the subjective method upon which we have already commented, but still they obscure a distinction which must be apparent to scientific analysis. We cannot know feeling except through realization, yet the knowing is not the realization. Being aware of the pain and the feeling pain are distinct acts of consciousness. All feeling, pain and pleasure, is direct consciousness, but knowledge of it is reflex, is consciousness of consciousness. The cognition of the pain as an object, a fact of consciousness, is surely a distinct act from the pain in consciousness, from the fact itself. The pain disturbance is one thing and the introspective act by which it is cognized quite another.

These two acts are not always associated, though they are commonly regarded as inseparable. It is a common 9postulate that if you have a pain you will know it, or notice it. If we feel pained, we always know it. This seemingly true statement comes of a confounding of terms. If I have a pain, I must, indeed, be aware of it, know it, in the sense that it must be in consciousness; but this makes, aware of pain, and knowing pain, such very general phrases as to equal experience of pain or having pain. But there is no knowledge in pain itself, nor pain in the knowing act *per se*. The knowing the pain must be different from the pain itself, and is not always a necessary sequent. We may experience pain without cognizing it as such. When drowsy in bed I may feel pain of my foot being "asleep," but not know it as a mental fact. We may believe, indeed, that pain often rises and subsides in consciousness without our being cognizant of it, but, of course, in the nature of the case there is no direct proof, for proof implies cognizance of fact. Pain as mental fact, an object for consciousness, not an experience in consciousness, is what is properly meant by knowing pain. Consciousness-of-pain as knowledge of it is not always involved by pain-in-consciousness as experience of it. Consciousness of pain by its double meaning as cognizance of pain and experience of pain leads easily to obscurity of thought upon this subject. But experience does not, if we may trust the general law of evolution from simple to complex, at the first contain consciousness of experience. This latter element is but gradually built up into experience, though in the end they are so permanently united in developed ego life that it is difficult to perceive their distinctness and independence. That pain and pleasure are cognized as facts of consciousness seems to us clear, but this does not deny that for us, at least, they may be cognizable only in fusion with other elements, as with sensation or volition. But whether known only with other elements or not, pleasure-pain is equally known only by direct introspection. I know directly and immediately 10pain and pleasure when I experience them, though they always occur bound up with some sensation. It may be that I never experience mere pain but some kind of pain, as a pricking pain, burning pain, etc., and that I always recall pain by its sensation tone, that I cannot isolate it by any act of attention. (E. B. Titchener, *Philosophical Review*, vol. iii., p. 431.) However I know that I have pain as well as I know that I have a pricking or burning sensation. "Did you feel the prick?" "Yes." "Was it painful or pleasurable?" "Pleasurable"; such a common colloquy implies as direct consciousness of the pleasure-pain as of the sensation. That I can at once discriminate a sensation as either pleasurable or painful certainly shows a direct awareness of pleasure-pain.

If pure pleasure-pain is primitive consciousness (see chap. ii.), it must be most rare phenomenon in such an advanced consciousness as that of the human adult: and it is not surprising that one should search for it in vain. But in any case it could not yield to attention. Attention as cognition views its object in relation, in a *milieu*; it can reproduce only by fastening upon something to reproduce by, but pure pleasure-pain has nothing connected with it. Again, attention as volition cannot reproduce mere pleasure-pain which is not volitional in its origin and growth like sensing, perceiving, or ideating. We merely "suffer" pain. Both pleasure and pain in themselves are purely passive; willing cannot directly affect them, and they are not, like cognitions, modes of volition, or effortful activities. For man to have a primitive consciousness by exercise of will would be quite as difficult as to turn himself into a protozoön.

Further, would not attention as introspective alertness to discover such a fact of consciousness as pure pleasure-pain denote that consciousness is thereby raised far above the level at which such a phenomenon can occur? In general also constant introspective attention tends to defeat 11 itself. A continual intentness and watching for a given psychic phenomenon is a state which, the more intense and persistent it is, tends to bar out the particular state watched for, and, indeed, all other states than itself. If attention as act engrosses, it defeats itself.

If, however, undifferentiated pleasure-pain should at any time occur in human consciousness, might we become immediately and spontaneously aware of it? By its very nature it may escape conscious attentive investigation, but may there not be a direct and simple awareness or apperception of it? We might suppose that one man tells another, "I was very sick, and in state of coma I had pain, merely pain, not any kind of pain or pain anywhere, but just pain, that was all the consciousness I had." Such an expression is intelligible, and may be a fact. However, it is in the phenomena of lapse and rise of consciousness that we see evidences that undifferentiated feeling probably occurs, and that sometimes in high psychisms. In the following chapter we discuss then this point as a matter of judgment of tendencies, rather than on basis of direct evidence of introspection, though this is not barred out.

CHAPTER II

ON PRIMITIVE CONSCIOUSNESS

Science views the world as an assemblage of objects having mutual relations. In this cosmos of interacting elements certain objects become endowed with mental powers by which they accomplish self-conservation. Just what these objects are and how they attain mental quality is beyond our direct investigation. However, assuming consciousness as a purely biological function, as a mode for securing favourable reactions, we can discuss the probable course of its evolution under the law of self-conservation. Mind, like all other vital function, must originate in some very simple and elementary form as demanded at some critical moment for the preservation of the organism. It is tolerably obvious that this could not be any objective consciousness, any cognitive act, like pure sensation, for this has no immediate value for life. It was not as awareness of object or in any discriminating activity that mind originated, for mere apprehension would not serve the being more than the property of reflection the mirror. The demand of the organism is for that which will accomplish immediate movement to the place of safety. The stone pressed upon by a heavy weight does not react at once to secure itself, but is crushed out of its identity; but the organism reacts at once through pain. It is certainly more consonant with the general law of evolution that mind start thus in pure subjective act rather than in mere objective acts, like bits of presentation or a manifold of sense. We shall now 13endeavour to elucidate this conception of pure pain as primitive mind, first from the general point of view of the law of self-conservation, and secondly from particular inductive considerations.

It is very difficult to conceive what this bare undifferentiated pain as original conscious act was, it being so foreign to our own mental acts. Our psychoses have a certain connection one with the other, and a connection which is cognized as such, so that the whole of mental life is pervaded by an ego-sense. But primitive consciousness must have been by intermittent and isolated flashes. The primitive pain, moreover, was not a pain in any particular kind, but wholly undifferentiated or bare pain. There was no sense of the painful, but only pure pain. Nor was there any consciousness of the pain, any knowledge or apperception of it. The pain stands alone and entirely by itself, and constituting by itself a genus.

Now to assert that this general pain exists, is not, of course, realism. The pain is a particular act, though it is wholly without particular quality. It is not a pain as one of a kind distinct from other kinds, but it is comparable to a formless, unorganized mass of protoplasm which has in it potency of future development. Pain may exist as such, but not a consciousness or a feeling. It is meaningless to say that the first psychosis may have been a consciousness in general form which was neither a feeling, a will, or a cognition, but the undifferentiated basis of these, nor can a feeling *per se* exist. The expressions, painful consciousness, and painful feeling are deceptive; there is no consciousness which pains, but consciousness is the pain, and the feeling is not pleasurable or painful, but is the pleasure or pain. "Feeling," as I have said (*Mind*, vol. xiii., p. 244), "has no independent being apart from the attributes which in common usage are attached to it, nor is there any general act of consciousness with which these properties are to be connected."

14Further, the law of conservation requires us to associate with this primitive act of blind, formless pain the will act of struggle and effort which is as simple and undifferentiated as the feeling. And these two we must mark as the original elements of all mental life. Strenuousness through and by pain is primal and is simplest force which can conduce to self-preservation. It is thus that active beings with a value in and for themselves are constituted. The earliest conscious response to outward things is purely central and has no cognitive value. The first consciousness was a flash of pain, of small intensity, yet sufficient to awaken struggle and preserve life.

Pleasure, then, we have excluded from playing any *rôle* in absolutely primitive consciousness. Pleasure and pain could not both be primitive functions, and of the two pain is fundamental in that the earliest function of consciousness must be purely monitory. Pain alone fulfils primitive demands, and secures struggle which ends in the abatement of pain through change of environment or otherwise. Pain lessens, but pleasure does not come, but unconsciousness instead, for no continuous organic psychic life is yet evolved. As long as pain continues there is effort and self-conserving action; when pain ceases, consciousness ceases, because the need for it is gone. Each fit of pain subsides into unconsciousness as struggle succeeds, and there is no room for even the pleasure of relief, which, indeed, must be accounted a tolerably late feeling. As far as the lowest organisms have a conscious life it is a pain life, but they have a Nirvana in a real unconsciousness. The evolution of pleasure must be accounted a distinct problem.

The law of evolution is, that origin of function and all progressive modification arise at critical stages. Thus it is in painful circumstances that the origin of mind is to be traced, and the important steps in its development have been achieved in severest struggle and acutest pain at 15critical periods. Pleasure is not then the original stimulant of will, but is a secondary form. Pleasure has an obvious utility which is far from the absolutely primitive. The pleasure-mode early enters, however, to sharpen by contrast the pain-mode, and it is only by their interaction that any high grade of psychic life could be built up. The development of pleasure cannot be from pain, but as a polar opposite to it. We cannot bring the development of mind into a perfectly continuous evolution from a single germ, as is the case in biological evolution. In a sense we may say that pleasure and pain are complementary, like positive and negative electricity, but the comparison cannot be pressed. We cannot, indeed, carry it so far as to believe either absolutely essential to the other. We mention, then, the evolution of pleasure as a problem which is yet to be dealt with in full. However, that it is not original element in mind is easily seen from this. As we ascend the grades of psychic life the pleasure-pain gamut lengthens, and as we descend, it shortens, with pleasure always as the intermediate factor. Thus, if we can represent it by a line,

Pain   Pleasure   Pain

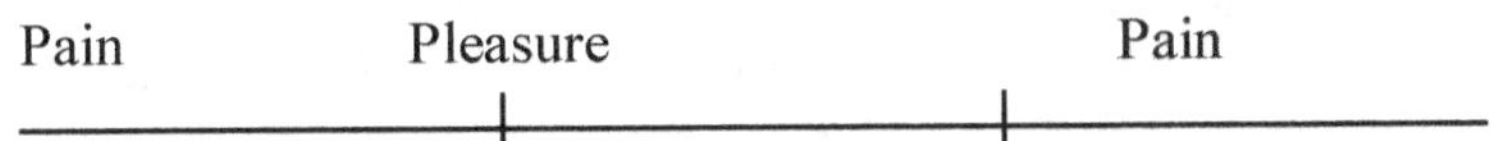

any single element which can affect psychic life, as temperature, moves through a highest pain intensity, an intermediate region, then to pain again as effects in a range from a very high temperature to very low, or *vice versâ*. Now, this gamut in a human being, from the intensest agony from heat to the greatest suffering from cold, consists of very many notes, but the step to unconsciousness is always at one end of the scale. In lower psychic life it shortens, but always at the intermediate points where pain merges into pleasure and pleasure into pain, and thus in the lowest form the original element of consciousness as feeling is seen when only the two 16extremes remain, namely, primitive consciousness as pain reaction. As the step from feeling—consciousness to unconsciousness is through a pain, this certainly points to pain as the original feeling, and the first element of consciousness. We must suppose then that the first organism which attained consciousness felt pain, that if this came from temperature, for example, that intense heat and intense cold would both produce a pain one and the same in nature, bare pain, not sensation of heat or cold. And this pain-consciousness response came at first only at the application of these critical temperatures, all other degrees not bringing any response. If consciousness like other functions originated as an infinitesimal germ at some crisis in life, it must have been with pain. The pleasure function, unlike the pain, does not originate in life and death crises.

That pleasure is secondary is also suggested by this, that pleasure is mainly connected with such late formations as the special senses, whereas pain is prominent with earlier functions. Thus we have pleasures of taste, but visceral pleasure is scarcely noticeable, though visceral pain, as colic, may be very acute. Wild animals, which feed often under fear of interruption or in extreme hunger, bolt their food without tasting, and so miss taste pleasure, and this seems to be the type of primitive feeding.

The origin of pleasure is then, I think, to be traced as an intermediary feeling between pain as produced by excess, and pain from lack as differentiated form. Pain as original and undifferentiated is the same whether resulting from excess or lack, but it is only after it has differentiated so far as to be in two modes that pleasure can enter as a mediate form of feeling and become a directing force to advantageous action. The primitive pleasure-pain gamut was this:

Lack Pain   Pure Pleasure   Excess Pain

—————————————————————————|————————————————|———————————————

17A general survey from the point of view of self-conservation leads us then to regard the original psychic state as a pain-effort form. There is first a purely undifferentiated sense of pain and closely consequent a purely undifferentiated *nisus*. There is neither sense of objectivity in general, nor in any special mode, nor is there feeling of pleasure. And the study of what seem to be the earliest forms of mental life in the child and in the lower animals points toward this conclusion. Preyer, in his studies on the mind of the child, expresses his conviction that the feelings "are the first of all psychical events to appear with definiteness," and that at first in no manifold forms. He adds, "The first period of human life belongs to the least agreeable, inasmuch as not only the number of enjoyments is small, but the capacity for enjoyment is small likewise, and the unpleasant feelings predominate until sleep interrupts them" (*Mind of the Child*, Part I., New York, 1888, p. 143, *cf.* p. 185). Since in the embryology of the mind as in that of the body the individual repeats in condensed manner the evolution of life, we judge that these observations point toward the genesis of consciousness in a single feeling state, pure undifferentiated pain. The earliest consciousness we can discover seems to approach this type. The close observer of very young infants must feel that the meagre psychic life they may have consists mainly of intermittent pains interrupted by comparatively long periods of unconsciousness in sleep. Of course, the earliest psychic life of the infant is not absolutely primitive both on account of heredity and on account of pre-natal experience; but in its general form it, no doubt, reverts toward the original *status* of mind. This original state, to which that of a very young infant is akin, was merely pain, which knew not itself nor its relation to other states, nor its relation to the external world, but was a wholly central subjective fact, and so was expressed only in wild and blind general movements. The very lowest types of 18psychic life which we can interpret seems to feel and nothing more. They do not feel *at* anything, and do not feel because they know, nor do they have definite kinds of feeling.

Pure feeling as bare pain and as undifferentiated pleasure is certainly far removed from our ordinary conscious experience, yet it may sometimes appear in a survival form, especially in sluggish states, in waking from sleep, and in recovering from anæsthetics. We are sometimes awakened by a dull pain which was evidently in its inception mere bare pain without differentiation. But in all such cases the pure pain or pure pleasure is but momentary, and is quickly swallowed up in a flood of manifold sensations. Many objects by many modes of sense at once invade and possess consciousness, and the early indefinite mode vanishes so quickly that we very rarely have time to note it by reflective consciousness.

But it is not merely in exceptional states of developed consciousness that we may trace the elementary form of feeling, but we may believe it to be fundamental to consciousness in general. It is natural for us who are so pervaded and dominated by sense of objectivity to see in it the causal element in mentality; feeling and will seem consequent to it, and we apprehend and feel accordingly. But the order of evolution was not from knowledge in any form to feeling, but the reverse, and we may suspect that in the completest analysis consciousness will still be found to obey its original law. If the rise of knowledge was at the instance of feeling, it is certainly unlikely that a fundamental order should be more than apparently reversed.

The order of consciousness is really the reverse of the order conceived by the objectifying consciousness, and this is a point where cognition by its very nature as objective may be said to obscure itself. To apprehend is to bring into relation, and the relation is very easily attributed to what is purely unrelated, to pure subjectivity. Thus here 19in the interpretation of merely subjective facts knowledge tends to stand in its own way. It is only objectively that the objectifying can appear causative of feeling; subjectively sense of object must always be taken as subsequent to a pleasure-pain psychosis. The object communicates or causes the feeling, but the subjective order is as such of necessity the opposite; the object does not come in view; there is no relating, until feeling has incited to it, and gradually the mind reaches out to an objective order from the purely central fact. In every psychical reaction there must be the purely central disturbance before the rebound to the actuality occasioning the disturbance. I must feel before I can discriminate or have any sense of the communication of the feeling. This means that when external objects are brought into relation with a wholly unanticipating consciousness, the first element in psychosis is always pure pleasure or pure pain. Thus, on a cold, dark day a sudden rush of sunlight on a blindfold man causes pleasure, then feeling warm, and then sense of warming object. The glow of pleasure and the pang of pain merely as such is in all cases precedent to any objective reference. Pure centrality of response, I thus take to be the initial element of all psychosis, primitive or developed. The first tendency in every consciousness is pure pain-pleasure, complete subjectivity which, however, in higher consciousness is so quickly lost through practically consentaneous differentiation that all traces of it seem wholly extinguished. Pure subjectivity must be pronounced the most evanescent of all characters in developed minds and yet the most constant. It is the inevitable precedent in every sensation and in every perception. We always experience pleasure or pain before the pleasurable or painful. A bright colour gives pleasure before we see it, and this pleasure incites to the seeing it. But so fully has the objective order been wrought into consciousness as a mode of interpretation that the great majority on reading 20the preceding sentence will mentally at first attribute sense of objectivity from the expression "bright colour gives pleasure," as if there were pleasure at colour, a colour-pleasure, whereas is meant pleasure and nothing more,—bare, undifferentiated pleasure.

The objective statement, however true, is no measure of subjective fact, but this twisting of subjective fact to correspond with objective order is so embedded in language and common thought that it will perhaps always remain the form of ordinary thinking, like common-sense realism and geocentric appearance. The expressions, it pleased me, it pained me, and the common modes of speech in general, are fundamentally misleading. Pleasure and pain bring their objects, not objects pleasures and pains. Pleasure *per se* does not come for and in consciousness from the object,—though this is objective order—but the object for and in consciousness comes from the pleasure. Pleasure and pain always precede any cognizance of the thing, and it is only the combination of the two elements that constitutes pleasure or pain of or at a thing. The primitive element, the original feeling movement, also excludes subject as real object; both the "it" and "me" are not yet apparent; there is not yet identification of experience with subject or object, and in fact no sense of experience at all. The psychologist must retain common expressions, however, but, like the astronomer who retains such phrases as the sun rises, the sun sets, he must reverse common interpretation and correct natural error.

Guided by this principle we note an obvious error in the interpretation of child consciousness. If a bright-coloured object is passed before the eyes of a young infant we may conclude from its expression that a pleasure-consciousness is awakened, but we are probably quite at fault if we conceive it to have a consciousness of bright, and that this consciousness preceded and gave rise to pleasure and gave it a *quale* as pleasure-brightness. Sense of pleasure-object 21is manifested by appropriative activities, but in the very young, where these activities are lacking, the response to object is best regarded not as in any wise sense of object, nor even any kind of sensation, but as a pure subjectivity of pleasure. Of course the same remarks apply to the pain side of the child's experience.

The purely subjective experience, while it becomes more and more evanescent factor as mind develops, yet always maintains its place as the initial point and vanishing-point of every psychosis. Every psychosis beyond the most primitive must be accounted a feeling-will-knowing group. These psychic forces exist in a correlated union generally comparable with the correlated activity of physical forces like electricity and heat. Each psychosis repeats in itself, in tendency form at least, the essential stages in the evolution of consciousness. Every psychosis rises from the pure pleasure-pain as the lowest level of mentality like a wave, and like a wave falls back into it again. Every wave of consciousness, whether it rises slowly or rapidly, whether it subsides gradually or violently, rises from pure subjectivity and comes back to it again. This absolutely simple feeling phase is accomplished so rapidly in ordinary human consciousness as to be rarely perceptible, but in lower consciousness it often exists as mood, as more or less permanent psychosis. The Brahmans attain artificially a subjectivity akin to this through their expertness in mental control and manipulation. They succeed in reducing and keeping consciousness in some very simple type, and their Nirvana may be considered as a state of pure subjectivity on the pleasure side. They, of course, cannot really attain this state or, at least, keep it, for pleasure is at bottom relative, yet they come to something approaching it. Pain at its height just before unconsciousness is reached, is always of the pure subjective type. In slow torture pain increases to a maximum intensity in pure pain, beyond which there is a gradual loss of intensity 22and consciousness in general, till ultimate failure of all consciousness. From the maximum intensity on to the end, consciousness is entirely subjective. Pleasure at its maximum attains only comparative subjectivity. Such facts tend toward a theory of mind which makes its original and fundamental act purely central; mind starts as in a germ which pushes outward till it penetrates space and time, but not in any reverse motion a pushing inward of a series of presentation forms.

We shall now notice certain of Mr. James Ward's statements on primordial mind—in the article Psychology, *Encyclopædia Britannica*—in which he controverts feeling as original and simplest unit in mentality. Mr. Ward regards *"the simplest form of psychical life"* as involving *"qualitatively distinguishable presentations which are the occasions of the feeling."* Presentation is primitive and initial in all consciousness, and cognition—feeling—will is the order for all mind. We always act as we are pleased or pained with the "changes in our sensations, thoughts, or circumstances" of which we are aware. Some presentation form is, throughout all our experience, the precursor and cause of feeling, and feeling can never be said to exist in a pure state as bare pleasure and pain totally without cognitive value.

On the contrary, I conclude from general considerations and from special indications in our own minds that pure pain is the original element, and that pure pleasure and pain are fundamental in all mind. Pure feeling arises from objects, indeed, but is still wholly unknowing of object and without qualitative aspect. Pure feeling is the constant incentive to all knowing and will activity. To say that I am pleased with a thing is to transform objective order into subjective fact. Pleasures and pains certainly come from things but this does not invariably rouse cognition of them as so coming, or of object as causative agent. The governing and essential fact of mind is always 23pure feeling, which, by reason of its perfect centrality, necessarily and naturally tends to elude observation. Every act of consciousness begins and ends with pure feeling, but mind, as far as it minds itself, is most apt to see only culminating phases rather than the obscure and inner forces which constituted long outgrown stages. The prominent facts of late consciousness are always very complex. Cognition as revealer unites with the known and inevitably, but strongly tends to regard itself as the determining and causative agent, whereas by its essence and function it is secondary. Cognition does not create its object, except in the view of a transcendental philosophy.

Mr. Ward asserts that phenomena of pleasure and pain involve change in consciousness with consciousness of change whereby we are pleased or pained. A changing presentation *continuum* is impressed upon mind, and it is by awareness of these changes that feelings are caused. This is certainly a complex mode to be assigned to all consciousness. This asserts that primarily consciousness merely happens in presentation form as determined from without, but I take it that the evolution of faculty is always acquirement, not mind determined, but mind determining, achieving its own growth in blind struggle. Mind is wholly an inward growth, not a series of givens; and presentations are accomplished not merely in it but by it. The fundamental principle is that while objects do determine conscious functions, it is only through self-conservative interest, through pleasure and pain reacting to them. All sensations, intuitions, presentations, are at bottom achievements as forced by law of struggle for existence. They do, indeed, seem to come of necessity and spontaneously to adult human consciousness, but developed faculty by virtue of being such does not have to attain beginnings.

But we note also this, that while all consciousness is 24change in the sense of being dynamic, of being an activity, this does not include consciousness of change. Consciousness as a changing factor is very distinct from consciousness of that change, and does not necessarily include or imply it. That the forms of activity which we group under the general term consciousness have their existence wholly in movement and change is true, but this does not necessitate that the changing elements should be aware of the change as such. Different things may be felt and known, but this does not always result in being known *as different*. This brings in comparison, consciousness of relation, which is certainly beyond primitive consciousness. In early mind we conceive that new elements are continually taking the place of the old, that change is incessant, yet without sense of the change. So far as the earliest consciousness is spasmodic and intermittent, appearing in isolated flashes, we cannot speak even of change in consciousness, much less of consciousness of change, for there is no continuous thread, no integration, consequently change is not in consciousness from a consciousness to a consciousness, but the only change is from a consciousness to unconsciousness. In the whole life of some organisms we may believe that only three or four pains or pleasures occur, entirely subjective and undifferentiated, and this collection of consciousnesses where state does not follow and influence state, where there is no complexity, is scarcely to be termed a consciousness which changes, much less that is aware of change. It is not improbable that even with civilized and educated men mind may sometimes lapse so far that changes occur with no awareness of change. In such sluggish conditions as when half asleep we may experience succession of consciousnesses without noting succession, each phase standing alone in itself and by itself. While consciousness is maintained as consciousness—that is, a continuance of conscious states—by the change, it is obviously not necessary to this that there 25should be awareness of change. Here as elsewhere we must keep clear of the mistake of making consciousness more than a general term for a group of phenomena. Consciousness as such has no reality or existence, but merely denominates a sum of consciousnesses. The phrase, change of consciousness, and similar expressions easily convey the impression that consciousness is a changing something. But we know that consciousness does not exist as a general indefinite something which changes or has other properties, but is merely a name for certain activities and functions.

The formula of Mr. Ward's hardly applies to developed consciousness, much less to undeveloped. Consciousness even in man cannot be regarded as a something which changes in sensation and presentation forms as pure givens, determined with immediate completeness from without, and these changes perceived, and pleasure and pain result. On the contrary the immediateness and spontaneity of presentation forms in our ordinary adult human consciousness are in appearance only; they stand first before us because they have reached a dominance through heredity and education, but still the latent and inward order is always from feeling to knowledge and not *vice versâ*. The accomplishment of presentation is usually so marvellously rapid in perceptive beings, and acts upon such slight incentive that it is only under very rare conditions of regression, or when developing a new sense or new form of sense that we see that the moving element in mentality is pure feeling. Thus, for example, in being awakened from sound sleep by a bright light suddenly brought into the room, the order of consciousness is, pure feeling of pain, sensation of light, perception of lighted object, and not the reverse; whenever we can catch consciousness gradually awakening we can always identify this order. The lighted lamp, objectively speaking, certainly caused the feeling of discomfort with which consciousness 26began, and this feeling roused the mind to both sensation of light and perception of lamp. I, of course, have a feeling as to the visible object only after seeing it, but this is altogether distinct from the feeling which incites to the seeing. A vague, undifferentiated pain or pleasure is always initiative, but pure pleasure-pain is often so low in intensity that it does not start any cognitive act.

In a general way the influence of feeling and emotion upon cognitive act in higher psychical life is acknowledged by common observation. The wish is father to the thought—we see what we want to see. What we observe depends upon prepossession, interest, and the whole pleasure-pain tone. The mind must be determined to cognitive act by interest of some kind, and even for advanced consciousness with all its strength of inherited aptitude total loss of interest ultimately leads to loss of perceptive power. The *impetus* of all previous cognitive effort will carry on cognition, of any high order, at least, but a comparatively short time. Blot feeling out of life and all nature would soon become a dumb show and quickly fade into nothingness. Absolute passionless receptivity is impossible under the conditions of reality, and pure presentation forms never *come* as antecedent and causative to feeling. We have constantly to bear in mind that in the nature of the case the simplest elements and fundamental laws are hidden and certainly far from conspicuous in highly developed mind, which is an intricate *nexus* of feeling, will, and cognition constantly acting and reacting on each other.

As a general statement, then, impliedly as to mind in general, and implicitly as to the developed human mind, the proposition that consciousness is fundamentally aware of changes in itself as the basis and cause of all feeling is an assertion which may well be questioned. Certain it is that being "pleased or pained with the change" is not feeling in general, but a particular kind of feeling, namely, 27feeling of variety and novelty. Further, to be pleased with a thing for itself alone is not to be referred to pleasure or pain "with the change." There is intrinsic pleasurableness and painfulness which does not come under the head of pleasure or pain of change. From both an *a priori* point of view of the law of self-conservation, and also from a brief survey of certain forms in comparative and human psychology, we incline towards accepting pure pain as the original consciousness which is very soon differentiated into excess and lack pain with evolution of pure pleasure. Will exists throughout as incited by feeling. Much, indeed, is to be done before this theory of the nature of mind is either fully elucidated or proved; but I believe that the assumption of mind as life function leads toward such a theory. Sensationalism and intuitionalism are both mistaken as to the origin and essence of mentality. Consciousness is not at bottom any mode of cognition, either as more or less freely accomplished by a "mind," or as more or less mechanical impression from "things," but it is primitively and fundamentally pain and pleasure as serving the organism in the struggle for existence. It is strange that evolutionary psychologists have so generally missed this point of view, and maintain sensationalism.

Comte, indeed, acutely remarks (*Positive Philosophy*, vol. 1, p. 463) that "daily experience shows that the affections, the propensities, the passions, are the great springs of human life; and that, so far from resulting from intelligence, their spontaneous and independent impulse is indispensable to the first awakening and continuous development of the various intellectual faculties." He here assumes the introspection which he elsewhere denies as psychological method, and enunciates an important principle which he never carried out. Horwicz has made a survey of feeling as fundamental aspect of mind, but his discussion is physiological.

Our conclusions have been founded on general considerations 28and on the phenomena of growth of mind in general and particular. Another line of evidence would be decadent mind. Mental powers should decline and vanish in the reverse of the general order in which they arose; the order of disappearance should be the reverse of appearance, and if pain-pleasure be primitive, we should expect to find it both the first conscious element in infancy and the last in old age. The last stage of senility seems sensitive only to organic pleasures and pains. Further, old age does not so much seek pleasure as guard against pains, and this fact is in line with our treatment of pain as prior to pleasure and more fundamental than it. We may consider it likely that conscious life in the individual begins with a pain and ends with a pain. Senile psychology on this and other points is worthy of far more attention than it has received, for it is on the whole more accessible and trustworthy than infant psychology.

With regard to Mr. H. R. Marshall's remarks (*Philosophical Review*, vol. 1, p. 632), it is sufficient to say that I lay no great emphasis on either pain or pleasure being the first fact of consciousness; but my main contention is that the primitive facts of consciousness are of the pain-pleasure type. While I have noticed some considerations as implying pain to be the first consciousness phenomenon, yet I am satisfied that pain and pleasure are correlative and complementary, each implying the other. Further, I do not regard pain as "primal sense," but as primal fact. Pain is not in any wise a sense, and sense of pain can only mean capacity for pain, or actual pain experience.

Again, I do not, as Mr. Marshall implies, regard pain as the differentiating basis of subsequent evolution, but rather as mere *prius* and impetus, and hence I do not look for pain-pleasure to disappear with mental evolution, nor yet to mark divisions in "sensational phenomena"; but it will ever remain in representative forms, at least, as increasingly complex stimulant of all mental life.

29The objection urged by Höffding and others to the primitive nature of pure feeling is that we sense before we feel pain or pleasure; thus we have the sensation of touch before we feel the pain from contact with a hot stove; we feel the pin, then the pricking sensation, then the pain. This precedence has been measured by Beau and others.

But what is the significance of these well-recognised facts? Do they show that pain-pleasure originates always in sensation? What is the origin of tactile power? How and why was the first tactile effort made, if not at impulse of some pain-pleasure? When conscious life was at pre-tactile stage—before it had learned to touch—it had no pain from touch, but it had pain. We can scarcely deny that a pre-tactile stage exists, that all sensation was originally a sensing—an exertive act, that it did not *come*, but was *attained*; for all the growth of sensitive power in the race proceeds thus at present, and the law of present psychic development in this regard seems general. But it is pain-pleasure which forces all action; here is the impulse which brings exertion whether as sensing or otherwise. A doctrine of spontaneity is against the general law of development by struggle. It is certainly true that, standing with my back to the stove and inadvertently coming in contact, I, without any previous pain-pleasure impulse and without exertion, have sense of touch, then pain. But this spontaneity is not original factor; it is the result of inherited powers. When tactility has become a well-developed power and is handed down to descendants, then contact with things is immediately and spontaneously realized in the form of touch, which contact would originally have been unnoticed. That is, the severest condition—a red hot stove—would impress the lowest psychism only in terms of mere pain, and so result in general reactions of *minimum* service. The early psychism which is just in process of achieving sense of touch would have pain, and then with effort touch 30the object and thus attain some more special reaction of more particular service. But the tactile, like all sensing activity is anticipatory, it is a finder, an interpreter. Suppose I bring a very fine needle toward your eye, you may see it and avoid it; but suppose your eyes are shut the eye comes in contact with the needle, and you have sensation of touch; but you are sound asleep, then pricking sensation may wake you as needle proceeds deeper, but in profoundest sleep undefined pain may be the first consciousness to result. Now the needle might be so small as to be seen with great difficulty by the waking man, or invisible, or to be touched with great difficulty; but this stage of exertive action for the sense is only relative, and in the history of mind the very grossest forms were at one time only dimly seen by intensest effort, and lower still, touched only by intensest effort. Seeing originated in looking, and passive touch in active touch, as moved by interest or direct pleasure-pain. Now pain is not in the mere sight or touch, but is suggested by them. The whole order—seeing, touching, feeling prick, feeling pain—is the reverse of evolution order. The rational mode, then, of interpreting the origin of any sense, whether tactile, visual or other, is not by receptivity, but through struggle at critical stage when great pain is actual or imminent. Thus, if the conditions of life required the development of a special sense of magnetism, it would surely arise by strongest effort, as, indeed, all progress in special sensitiveness is now being accomplished. Thus, the anticipatory and premonitory function of sense does not make it original, rather the contrary; it is guide and significant of pain-pleasure.

It is obvious that the cognitive tendency once established becomes an instinct of objectivity and governs the whole mentality. This is obviously the case with man. He does not exist in that sluggishness and semi-consciousness where pain-pleasure must arise as primitive impulse, 31but by habit and instinct he is passively and actively cognitive. The eye is continually seeing things spontaneously, the hand touching, but as to some very small object we have to exert effort to see or touch, and this was undoubtedly the mode by which all seeing and touching arose. It is because generations of ancestors actively sensed, that we automatically sense; the tendency has become ingrained in mind. So it is that man is predominantly sensing, is continually and naturally awake to objective conditions, is constantly anticipatory, and so normally senses before he feels pain-pleasure. However, a man in a "brown study," inadvertently touching a hot stove, has pain, then warmth, then touch sensation, and actively realizes these. So in deep slumber mentality often begins with pain-pleasure. At bottom the reason we have pain from a sensing is because we had originally pain-impulse to that sensing, and the pain therewith. Thus tactility, arising as effortful sensing, was produced by pain from thing to be touched, to be sensed in its experimental value. By innumerable painful experiences with hot things, the hot thing is tactilily appreciated; and as touching is actively pursued by organism on the alert, the associated pain is more and more quickly realized from given object. In origin pain was felt from the hot thing in contact, before either sense of warmth or contact was sensed; it was this pain that forced to sensing and development of cognition, which, however, ultimately became habit, and things were constantly appreciated and anticipated. Thus the touch-warmth-pain order is established. Sense is significant of pain-pleasure, but the pain-pleasure came not at first from the sensing, but the contrary; sensing was determined by it, and became correlated with it, and became sign of it. The progress is from initial subjectivity to an instinctive constant objectivity. This objectivity is reflected in all objective expression as language; "the heat was painful," "it hurt"; the "it" being 32tactual thing, etc., etc. However, if we look for primitive consciousness, we must find it only in primitive organisms in their primitive stage, and in man most rarely only as tendency in profound relapse. We must mark this, that cognition is not to be evolved out of feeling, but at instance of feeling as impelling the knowing effort or volition.

We may suppose that primitive consciousness still exists in the lowest types of life, but it may also be the sub-consciousness in the higher types. Viewed biologically, what is sub-consciousness?

The earliest living aggregations attain but a very slight degree of common life, and very slowly do the cells, under the pressure of serviceability in the struggle for existence, give up their independency and become interdependent, each thereby giving up some functioning to be done for it by others, and in turn functioning for others. Thus it is but slowly that a stomach is specialised, the cells in general in the organism long retaining and exercising some digestive function, which is properly termed sub-digestion. In this way a soup bath gives nourishment. If psychic function specializes gradually like other functions, we shall have in the same way a sub-form here, a sub-consciousness which stands for lower centres, and not for the whole organism as such. The wider, higher, and more specialized psychic centre does not at once extinguish the lower.

Now what is a *high* organism but an involved series of combinations of combinations? With every new integration a higher plane is achieved, and the vital process has a wider functioning: but the physical or psychical activity so far as it does not pass over into the service of the new and higher whole remains as sub-function. With every new stage in evolution the integrating psychic factors only partially lose themselves in effecting a common psychism for the new whole, a sub-consciousness and a sub-sub-consciousness, etc., are still carried on in survival. In man, 33physiologically speaking, it is the brain consciousness which is general. But we need not suppose this to extinguish all the lower ganglionic consciousness from which and by which it arose. If psychic function be correlative with other function, we must expect in man a vast amount of survival sub-mentality which, while not the mind of the man, is yet mind in the man. The individual knows necessarily only the general consciousness, for this only is *his* consciousness and constitutes his individuality, yet the doctrine of evolution would call for a vast deal of undiscoverable simple consciousness which never rises to the level of the whole organism's consciousness. A cell or a group of cells may be in pain and yet there be no pain in the individual's consciousness, and so unknown to this general consciousness.

We have intimated that primitive consciousness may occur in a sub-conscious way in the highest organisms. But can this sub-consciousness ever be more than mere survival in its nature? or may it play essential part as basis of higher manifestations? If the integration of mentality is like other integration,— *e.g.* material which is based on molecular and atomic activity—it will be bound up in the activity of psychic units, which can be none other than sub-consciousness. That is, any common or general consciousness when looked at from below, and analytically is the dynamic organic whole of elements; it is a product of activities which are on another plane from itself. Roughly illustrated, I may say that my finger feels pain before I do. We conceive that at a certain intensity a sub-consciousness tends to rouse a general consciousness, and for a time maintain it; and losing intensity, the general consciousness disappears leaving only the sub-consciousness, which may long outlast the general form.

Sub-consciousness, whether as survival or basal, is put beyond our direct observation, but it remains a necessary 34biological and psychological hypothesis. Here is exemplified for psychosis that law of the aggregation of units in hierarchical order, that wheel within wheel structure of the universe, upon which I have touched in *Mind*, ix. pp. 272-3.

CHAPTER III

THEORIES OF PLEASURE-PAIN

The bearing of our studies on a theory of the conditions of pleasure-pain is obvious. If we consider pure feeling as the primary, fundamental, and conditioning mentality, it stands before all other mentality, and cannot be interpreted as conditioned. Pain as *primum mobile* is not intrinsically dependent on any other psychosis. Hence we run counter to the Herbartian School, which maintains that psychism exists from the first for itself as intellectual ideational activity, and that pleasure-pain is but reflex of the efficiency and ease, or the inefficiency and difficulty of this activity. The checking of the current of ideas may give a pain, but our exposition has been that pain arose before ideas or presentations of any kind, and long before any interference could be felt as pain.

Again, if we say "all pain comes from tension" (*Mind*, xii. p. 6), we have to ask, Tension of what? If we say tension of sensation or ideation, this is Herbartianism merely. How also can tension be felt as painful, except through sensation of tension, which is a feeling of intense sensation—obviously a late psychosis? And certainly pain is more than a general consciousness fatigue. And further stress and strain result in pain, because we imply these as painful activities by the very notion of the words. A stress or strain is assumedly painful activity, but this is not explanation. But apart from this, if the organism felt pain merely as direct result of struggling and straining, it 36would cease activity; activity and evolution would stop. It may be that by tension is not meant a mode of consciousness, but of nervous or muscular activity; but as we are now considering psychosis only as conditioning pure feeling, we leave this aspect for discussion till a little later. But on the psychical side, that all pain is a by-product of over-intense consciousness, intellectual or volitional, that the origin and development of pain is in a mental intensity which has gone beyond a certain point, this seems, on general evolutionary grounds, unlikely. Here, indeed, is merely a very particular and rather late mode of pain. And may not pains themselves attain an intensity which is itself painful? It must be acknowledged, however, that the whole doctrine as to consciousness intensity, its nature, reactions, laws, and measurements is very obscure.

Again, as to the theory that pleasure-pain is reflex of quantity of consciousness, that pleasure results from mental expansion, pain from mental contraction, this must, like the intensity theory, be considered as putting a late and special form as covering all forms. Mentality here exists for itself, and conscious self-development—a very late mode—is presupposed. The promotion of large complete free consciousness, the sense of progress and of unimpeded mental activity, certainly conveys high joys to certain choice natures, but they do not touch the vast majority of even human minds, much less animal. With the stolid an expanding consciousness is painful. Consciousness only as conscious of itself, and as self-developing, reaches a pleasure or pain as a felt furtherance or hindrance of its own expansion.

All reflex theories take us above the realm of simple consciousness acting directly for life, and this is the very form which seems commonest, and which appears to be full of passing pleasures and pains. That consciousness does react on itself in late phases is plain, but if consciousness, like other functions, has developed from the extremely 37simple to the extremely complex, this self-reaction cannot be regarded as primitive. Not till consciousness becomes integrated as a manifold organism do pleasure and pain become prominent as reflexes. We are not now looking for the functional value of pleasure and pain in mind itself as an independent whole; but regarding its functional quality and that of all mentality in life values, and here the functional meaning of such reflexes is secondary. In mind, as organic continuous whole, pleasure-pain is both resultant and excitant; it stands related to an antecedent state and it is stimulant to following states. Its function is excitant and it is the starting point of all other mentality, both originally and in the later manifestation. The having pleasure-pain is what starts both motor and cognitive volition.

It has, indeed, been maintained that while pleasure-pain is not a product or concomitant of some psychosis, as sensation, it is itself a sensation, a definite mode of sensibility. I have a pain sense just as I have a temperature sense, I feel pain in the same way as I feel warm, and by the analogous sensory nerves. With reference to this theory we must ask, since sensation is correspondent to modes of objects, to what mode is pain correspondent? Sense responds to modes of object, as light, and sonorous vibrations; but pain is not based on any such mode of objects. If pain were, there would have been long since a department of physics, which would have treated that basis just as it treats light, heat, sound, etc. But we all know that an object is not painful or pleasing in the same way that it is warm or cold, heavy or light. I do not say the stone feels heavy and painful, but I do say the stone feels painfully heavy, that is feeling pain is not a state of awareness. Further, having pain or pleasure is not by any sensing effort. I do not try to feel pain as I try to see the light of a star or feel the warm spot in a bar of 38iron. To be sure, the doctor asks his patient, "do you feel any pain?" and after a moment's delay the answer may be, "yes," but this is not in the nature of a sensing effort, but merely an attentiveness to bodily conditions as affecting mental state, not an objective attention but an analytical self-attention. Still further, a neural basis for pleasure-pain is altogether likely, but even if these nerves were found to be generally distributed over the body, this would not prove sensation, but merely that pleasure-pain is functional throughout the organism, diffusive organic consciousness. If pleasure-pain is primitive, and neurality and mentality correlate, the earliest nerve structure—ganglion—was a pleasure-pain organ. However, the sensory motor predominance is so early and complete that the current theory, as the more objective, is the natural physiologic interpretation.

Again, it has been maintained that pleasure-pain is not a definite state of consciousness, but a quality like intensity, a *modus* which must belong to all states. But if we assign pleasure-pain to such a category as intensity we must define just what we mean by this category. Is intensity a mere objective quality which we as observers assign to all psychosis, just as we do to electrical or luminous phenomena? or is it inherent element, an actual constituent, of every psychosis? If a man is angry and becomes more angry, intensity is increased; but we may conceive that he simply is more angry without being aware of this change of intensity, that is without every change of intensity being noted by consciousness. As introspection avers, it often happens that a man is both unconscious of his anger and unconscious of its increase. As I have frequently had occasion to note, simple natures are wholly unconscious of their emotions and of their intensity variations. That is, as matter of fact, intensity of feeling is not feeling of intensity. If you feel warm you feel differently than when you feel warmer, but this is 39no more than saying that when the iron is hot it is in a different state than when it is hotter. Intensity means the same in both cases. Consciousness, primitively, at least, is not self-awareness of its own changes in intensity. The feeling warm and the feeling warmer occur simply as facts which are subjectively unrelated and unmeasured by the consciousness which has the varying intensities. I strike a cow hard—result, intense pain; harder, more intense pain; this is correlative with, I strike iron, intense tremor; harder, more intense tremor. The cow experiences more intense pain, but does not consciously measure it off as such. I can say, "I feel hotter than I did," but the cow does not appreciate and express its own sense of its experience. The language fallacy leads us astray. By our very use of terms, warm and warmer, and by our discussion of the matter, we imply a consciousness of intensity which is far from being primitive or general. It would probably be an overestimate to say that the intensity of one in a thousand psychoses makes itself felt as such in consciousness.

That consciousness is not always conscious of its own intensity is then shown by direct introspection. And in general we must observe that every psychosis has its own intensity, which intensity may or may not be noted by a consciousness of intensity. If there come a consciousness of intensity, this consciousness has its own intensity, which may be noted by a new consciousness, whose intensity may in like manner be noted by a new consciousness, etc., *ad infinitum*. That is, a consciousness is never its own intensity, and intensity is never a consciousness, such as pain or pleasure, but is mere comparative objective quality.

Again, consciousness has almost from the first different degrees of activity, but it would be most unlikely that so complex an act as consciousness conscious of its own intensity should be primitive and early. Also, if consciousness 40develops as life factor it must be immediate utility which determines its early forms. Hence on this general principle of biologic evolution it is most unlikely that primitive organisms will both have consciousnesses and consciousness of their intensity, for of what direct and vital value is this intensity-consciousness as psychic mode? On the other hand it is obviously desirable that psychoses should early differentiate intensity as objective quality, *i.e.*, without self-awareness of it, should have different degrees of a psychosis to meet different degrees of requirement; thus to fear strongly or weakly according to necessity of the case. To have fear set at one pitch for all cases is perhaps absolutely primitive, but differentiation is early. But to fear more or less, *i.e.*, at different intensities, is not to have intensity as subjective element, an actual psychosis constituent appreciated as such, which is very late evolution since the demand for it is late. In thus defining the category of intensity we have plainly isolated it from the pleasure-pain category. We know pleasure or pain as act of consciousness just as we know volition or sensation. Pain and pleasure are definite facts like seeing or touching or willing, and are so recognised by common consciousness. One or the other may be involved in all experience, but this does not make them general qualities like intensity. Pain is a consciousness, intensity is not a consciousness. This is the immediate value of the terms, the very names convey distinctness of category. I have a pain, I do not have an intensity; I am in pain, I am not in intensity. My pain is intense, but I cannot say my intensity is painful. We experience pain and pleasure, but we never experience intensity.

This *quale* hypothesis as presented by Marshall in *Pain, Pleasure and Aesthetics*, is set upon the dangerous foundation of ignorance, viz., of the neural basis of pleasure-pain, and of causes of its variability. It is as yet disputed whether a nerve organ for pleasure-pain has 41been found; but if one is generally acknowledged, the theory would be overthrown. Greater intensity in any psychosis, as sensation of warmth, means simply greater nervous activity in the particular nerves subserving the psychosis, in this case the temperature nerves. So also pleasure-pain as general concomitant like intensity must mean merely some general mode of nervous activity as yet unknown, if we allow it any nervous basis at all. Again, the variability of pleasure-pain for a given content, the fact that the taste of olives is at one time pleasant, at another, unpleasant, suggests that pleasure-pain is like intensity merely a general quality, which must in one form or another attach to all psychoses. But this does not explain anything. What we want to know is why in any given case we have pleasure and not pain; we do not wish to be put off with a general statement that the nature of pleasure-pain is such that we may have either, which is akin to the old metaphysical method of abstract explanation; making the *rationale* of the lion leoninity is not unlike the hypothesis that explains pleasure-pain in all its variations by variability as its nature. We have a scientific faith that variability is not a general unexplainable quality, but that there is for every case of pleasure-pain a definite *rationale* based in the facts of life demand and life history. That olives now give pleasure, and now give pain, is based upon definite conditions of physical state which are very complex, but which can be revealed by patient research alone.

Any theory of pleasure-pain then from the point of view of pure psychology, as explaining it by reference to other modes of consciousness, is, we think, unsatisfactory. But perhaps the physiological point of view will be more satisfactory. It is generally considered that the function and origin of pain is in what is unfavourable to physiological function, of pleasure, in what is favourable. I cut my finger, and the pain says, stop the injurious action. However, 42there are exceptions. I taste sugar of lead; it is pleasant, and I keep on tasting, and am poisoned. Lotze explains that this sweetness is immediately soothing and advantageous. "We must not regard pleasure," says Grant Allen, as "prophetic." But what has been the evolution of taste as sensing act except to be "prophetic," to give at the opening of the alimentary canal a monitor to the stomach and other digestive organs? That it tastes sweet, that this taste is pleasant, and so the substance is swallowed, or that it tastes bitter and unpleasant, and the substance is rejected; this surely is anticipatory and "prophetic." The taste for sweetness is not evolved for itself; but for its life value; and hence Lotze's explanation fails from the point of view of evolutionary psychology. The organic sweet is the nutritious and beneficial, and the sensing this quality in connection with these favourable and pleasant effects on the stomach and organism as a whole has led to a taste and liking for sweetness. "Sweet and wholesome" is the common and just conception. But if mineral sweets injurious to life, like sugar of lead, had been a common environment, and the only sweet known, this sweetness would have been as unpleasant as the sour or acid now is. We see even now that sweets that have several times caused nausea, though at first highly agreeable, come to be distasteful and disgustful. We now find that sour and bitter substances are disliked by animals in general as painful, for the sour and bitter is general sign of the unwholesome; but those animals which live almost exclusively on bitter herbs undoubtedly appreciate this quality as we do a *bon bon*. Men lost in a desert by pertinaciously tasting bitter herbs and becoming dependent upon them for support would soon realize their bitterness as pleasant, and a race might originate to whom sweetness would be unpleasant. Hence the value of a sensation does not—in natural evolution—lie in itself, it is merely a guide and index; and the sensation quality will be pleasant or 43unpleasant according to its relation to the demands of life. A sensation is inherently either pleasurable or painful, but not essentially one and not the other, hence the proverb, *de gustibus non disputandum*. The sensing act in itself is indifferent, *i.e.*, sweetness and bitterness, purely as tastes, as sensing acts, are indifferent; but as matter of fact having grown up with and for pleasure-pain tones as indicative of life values, they are either one or the other according to their relation to life. Where sense serves not life but itself, as with the epicure, a new order of pleasures and pains is determined which is not within our present scope of discussion.

This variability of pleasure-pain tone of sensations even under natural evolution shows that the main force at least of their pleasurability or the contrary does not lie in the affection of the sense organ itself. If a given sensation, for example, bitterness, were painful in all degrees only because of its harmfulness to the sense organ, how could this variability be explained? We consider that the tasting bitterness, for example, arose through painful stomachic and bowel experience with herbs which had this quality, and which by sensing efforts were so cognized at length, and pain connected by its very origin with sense of bitterness, which becomes in all degrees painful. The identifying the nutritiously harmful weed by tasting its bitterness has the pain quality of its effects, since the tasting has grown up in connection with its effects. It is out of actual injurious and painful experiences that the organism is led to put out sensing effort and to reach such a sensation as that of a bitter taste whose pain value is mainly, at least, due to the actual results of the substance lower down in the alimentary canal. A sense of bitterness becomes disagreeable in all degrees, for in its inception, when first sensed, it has its connection with the pain effects which stimulate this sensing. To discriminate the unnutritious or poisonous by tasting is a grand achievement, securing 44the rejection at the very opening, the mouth of the alimentary canal, in place of rejection by nausea from the stomach itself. The organism which could only know that a certain substance was bad for it by very painful nausea, now knows its badness by the comparatively painless tasting bad. Whatever tastes bad, is bad.

The chief difficulty of the theory of bodily advantage and disadvantage as conditioning pleasure and pain comes not from any such instance as the sugar of lead phenomenon; but it lies in the fact that life progressiveness, enlargement, specialization, that which is to the highest profit of life, is uniformly reached only by painful struggle. It is only by intense struggle, by supremest, painfullest effort, that those new psychic forms are initiated and developed which are of the utmost service to the organism. The act of adjustment to a new circumstance is so extremely difficult and painful that it is attempted by few and achieved by very few of any set of organisms. By an act of most painful struggle the fittest survive; and the rest, the vast majority, who could not key themselves to that pitch, perish. Adjustment to the ordinary conditions is simply a free using of intelligence and energy integrated and stored by ancestors when these conditions were new to them. The adjustments which are so spontaneously made by new-born animals as response to environment were once new, and secured and integrated for inheritance by the most painful and persistent effort. Such is the inertia and conservatism of life that while it moves spontaneously in grooves already made, it does not rejoice in the toil of real progress. The struggle by which the greatest life advances have been accomplished has always been intensely painful in itself, whatever the aftermath of pleasure may be, the pleasure of achievement and creation, the satisfaction at successful effort, which is plainly a very late psychosis.

The origin and place of pleasure is indicated by these considerations. Though function is generated and 45developed by severest painfullest struggle, yet the reward is pleasurability of the free functional activity; and the more manifold the functioning built up, the more manifold the pleasure. Thus it is that a highly complex organism like man, which represents many psychic ages of painful function building, has a very high pleasure capacity. Every new adaptation when integrated means a new pleasure. It is pleasurable to inhale fresh, cool air, but the lung functioning itself has been built up by painful exertion in the struggle for existence. Pleasure as reflex of functioning is merely then conserving power. The immediately and intrinsically pleasure-giving acts are not progressive, but merely hold life at the given and already acquired status. But the most and largest pleasure is in the mere expenditure of stored energy. The easiest way, the way of inclination and obvious direct pleasure is regressive. It is living upon the past, living upon accumulated capital bequeathed, and perhaps in some measure acquired. The use of a stimulant, as alcohol, enables the capital to be used up faster. As the systemic craving becomes greater with the drunkard, the pleasure increases, and on the brink of dissolution he may reach the extremest pleasure. In alcoholism the more injurious the drink, the more violent the pleasure. The most rapid and destructive using up of vital force in lust, revenge and other excitements gives the keenest pleasure. The orgy, the chase, the prize ring, give the expensive "thrill," which is ecstatic pleasure. Debauchery and alcoholism are quick ways of using the pleasure capacity which has been built up by painful effort of thousands of generations. A taste sensation, which was achieved as the highest effort of genius by some very remote ancestor at a critical moment and attained by painful sensing exertion, is finally after generations of severe volition integrated, and becomes spontaneous activity, and reactive as free pleasurable functioning. That is, in the early stages of tasting the pleasure taken in it was by discriminating 46effort, a pleasure realized by exertion as pleasures of artistic "taste" are now enjoyed by many people; which pleasure may at length be so inwrought into psychism that it occurs spontaneously. At least, we have no other clue to the origin of pleasures except by judging from the present development of definite pleasures in the case of man, which pleasures come only by effortful cultivation, for instance, the highest pleasures of art. The whole range of sense pleasures have been built up and capacity therefore has been inherited, and may be used up with great intensity.

The largest and keenest sort of pleasures is from expenditure. Yet storage in certain modes yields a moderate pleasure, as the pleasure of rest, dozing after exercise. Here is a general spontaneous accumulation of physical pleasure capacity, it is a case where functional repair has become automatic, and thus far is analogous to the spontaneity of pleasures of expenditure. But these storage pleasures are mainly negative, relief only; and they are not the great positive corporeal pleasures which are so largely sought. The drunkard gradually recovering from a spree experiences feelings of relief, but he does not indulge in his cups to feel the gradual recovery from the painful after effects.

No biologic or psychologic theory of pleasure and pain can yet be enunciated which is fully explanatory. In fact, if pleasure-pain is the primitive and fundamental fact, if it constitutes the worth of life and is life, then it must explain other factors, but remain itself unexplained. The theory of advantage and disadvantage fails signally, for the most pleasurable act is frequently the most disadvantageous to the interests of the organism, and the most advantageous—progressive effortful volition—is invariably most painful. As to why the way of conservation and upbuilding should be painful, why pleasure should not be inherent in the progressive struggle rather than pain, is, at least for the 47present, a philosophical problem; but the fact remains. We have considered that struggle is pain-impelled and painful, and that pleasure is resultant of functioning thereby established, and that all pleasure capacity is painfully acquired. With the grand exception of this singular and important fact, however, we can say that in natural evolution—that is, before mind has become independent and artificial and subjected itself to pathologic tendencies—the general law that pleasure denotes favouring organic conditions, pain, unfavourable, may be assumed. However, if the body is mere dependency and expression of mind, the form of statement must be reversed; that is, a given pain or pleasure is an acquirement by mind in its function building. I have painful taste sensation of bitter, pleasant sensation of sweet, not as originally reflex of bodily conditions, but the sensing power and the organ, like all bodily specialization, is outcome of mind as struggle. A typical consciousness—series of a low type which places pleasure in its place is: pain (as from hunger)—struggle-sensing (as touching for food)—desire (when food is recognised through sensing)—absorptive and digestive effort and action—pleasure—struggle to continue and increase pleasure—slight satiety pain—unconsciousness of sleep. So we do not connect pleasure-pain as outcome of organic function in general or particular, but function is outcome of pleasure-pain. It determines function, and not function it. The feelings which prompted and developed a functioning, and the correlate total—organism—necessarily involve a very high complex, at least for any late psychism, and make a general law of pleasure-pain impossible to determine under present conditions. The *rationale* of particular pleasures and pains can only be reached through a thorough investigation of life history, an investigation which in present circumstances seems in most cases beyond our powers. A great mass of psychological *data*, and not any general theory, is the *desideratum*.

CHAPTER IV

THE RELATION OF FEELING TO PLEASURE-PAIN

Should the term Feeling be made to include certain states of consciousness which are neither pleasurable nor painful? Or should all such neutral states be designated by some other term? We are concerned here with an important matter of definition which implies an extensive analysis of consciousness with reference to pleasure and pain. It will not be difficult to find many so-called feelings which are neutral, or seem to be so; but it is the duty of the psychologist to carefully analyse all such states, and point out the proper use of the term Feeling.

Common observation neglects minute analysis, and is unreliable when it speaks of certain indifferent states as feelings. When a man speaks of feeling queer, or strange, or bewildered, or surprised, and says that the state of mind seemed neither agreeable nor disagreeable, we may suspect that by a perfectly natural tendency he is extending the name Feeling to closely-connected states of cognition or will. In identification and definition common observation is for all sciences notoriously untrustworthy, and especially in psychology; so on this question the evidence of language and popular testimony counts for little one way or the other. This is strikingly evident when people speak of feeling indifferent as to some matter, meaning that they have no feeling on the matter. The term Feeling is used in such a broad and vague way that 'I feel indifferent' means 'I am indifferent,' 'I have no feeling.' The mistake here is in using the word Feeling 49as an equivalent to Ego, or any quality of Ego. A feeling of indifference is no feeling at all. Popular evidence then, I believe, can be no guide in this matter. In passing, I may also say that the very abundant use of analogy by some writers on this subject seems to me ill-advised. Analogy does very well to bring up the rear, but it is often very useless and confusing as an advance-guard.

Prof. Bain (*Mind*, No. 53) insists that ideas tend to actualise themselves by neutral intensity or excitement, which is feeling; or rather, he says, a "facing-both-ways condition." This last expression is certainly not very helpful or satisfactory. Prof. Bain admits that typical will is incited by pleasure and pain, but he maintains that sometimes, as notably in imitation, will is stimulated by purely neutral excitement or feeling. In the discussion of this subject much has been said about excitement, and, as Mr. Sully has suggested, this requires careful definition.

Reflection assures us that every mental activity has a certain intensity, and the word Excitement may, in the most general sense, denote this intensity. The intensity may be so slight as to be unnoticed by the subject, and remain wholly unindicated to the keenest observer; or it may be so strong as to be perfectly evident to both; or it may be evident to the subject and not to the observer, or *vice versâ*. Thus the obvious division of Excitement from this point of view is into subjective, where it is immediately recognised and felt in the consciousness of the subject, and objective, where it is unnoticed, or noticed only by observer. Classifying by another principle, we may distinguish Cognition-intensity, Feeling-intensity and Will-intensity, and the natural subdivisions under these according to the accepted subdivisions of mental activities. Excitement is not, however, generally used in the large sense we have just mentioned, but as denoting intensity of a high degree so as to be very noticeable to the subject, or observer, or both.

50It is plain that Excitement, as subjective intensity, is the only kind which bears on the question under discussion. It is with excitement as a feeling, *viz.*, the feeling of intensity, and not with excitement as quality of feeling, that is, intensity, that we have to deal, and it is necessary that this distinction be clearly borne in mind. One may be excited but not feel excited, may have intensity of feeling but not feeling of intensity. Using the term, then, as equivalent to feeling of intensity, it is to be noted that it is a reflex or secondary mental state. It is the feeling resulting from consciousness of intensity of consciousness. The intensity of any consciousness may increase to such a point that it pushes itself into consciousness, first as mere recognition of intensity, but immediately and most manifestly as feeling of intensity. In rapid alternations of contrasted states, as of hope and fear, intensity soon rises to such a degree that it forces its way into consciousness as feeling of intensity. This feeling of intensity may be itself either weak or intense. In very reflective natures, the cognition and feeling of intensity may be reflex at any power: there may be cognition of the intensity of cognition-of-intensity, etc., in indefinite regression. Most persons stop with the single step in the regression.

It is evident that as far as excitement is regarded merely as intensity, as a fundamental element in all feeling and mental action, it is a confusion of terms to apply quality to it, to speak of it as either pleasurable, or painful, or neutral. Intensity of mental action has degrees but not quality, just as pitch in sound has degree, but not timbre or quality. Regarding excitement as feeling-of-intensity, it has the general characteristics of all feelings, and is not more likely to be neutral than any other feeling.

Taking the case of surprise, which is so frequently instanced as a neutral feeling, let us analyse it with special reference to the excitement as feeling of intensity of cognition. A typical case would be the surprise from 51hearing thunder in January. The presentation is quickly compared with a representation of observed order of facts, and the disagreement of the two marked. This is so far purely cognitive activity; but immediately connected with the perception of disagreement is the forcible recognition of the breaking up of a more or less rigid order. There is a disturbance in cognitive activity and the tension breaks into consciousness as excitement, the feeling of intensity. The conflict of a settled conviction with recent presentation intensifies consciousness, and this intensity with the abrupt change in quantity and quality of mental activity breaks into consciousness as intellectual sense of shock accompanied and closely followed by feeling of unpleasantness and pain. It is to be noted that when we come upon the feeling-element in surprise we find pain. Surprise in the strict sense is then the reflex act of consciousness in which the mind becomes aware of and feels the sudden disturbance and tension set up in itself by the sudden weakening of an established belief. The painful shock has some relation to the force of the disturbing factor, but is more closely connected with the strength of the belief assailed. The feeling of the disagreement as pain is due to the fact that this disagreement impinges on subjectivity, personal opinion and conviction, and the disturbance will be more or less disagreeable according to the degree of personal interest. Note that by exact statement the feeling is not painful, but is the pain concomitant or resultant upon the mental perception. The surprise for a person of rather weak habit of mind and of little generalising power will be almost wholly intellectual. Disagreement will be noted, but not felt. For one of strong intellectual interest, the surprise will mean definite and acute pain. For a meteorologist who has written a book stating that in this latitude thunder does not occur in January the surprise might be very grievous. The intellectual element in surprise is emphasized in the 52statement "I am surprised," the feeling-element in "I feel surprised." If antecedent states of representation, comparison and inner perception are placed under the term feeling-of-surprise, we may expect consequent states to be likewise easily confused. When one speaks of being agreeably or disagreeably surprised, the pleasure or pain is not really, however, a part of the surprise. The sense and feeling of intellectual destruction, which constitutes surprise, is so quickly and thoroughly swallowed up in pleasure in having hope realized, or in pain in having fear realized, as the event may prove, that the term is naturally applied to what engrosses attention. Thus, "It was a very pleasant surprise" means "The surprise was followed by very pleasant consequences." When I am surprised by the arrival of an intimate friend whom I supposed a thousand miles away, the mental disagreement, and the pain from conflict of conception and perception, are quickly eliminated by the event according with desire, and by the mind anticipating joys. We see, then, how easily the antecedents and consequents of surprise are confounded with surprise itself, which is the reflex act of consciousness recognising and feeling sudden disturbance in intensity, quality and quantity in cognitive activity. I conclude that surprise, as feeling, is pain coloured by cognition of shock and by volition to avoid disturbing element.

Absorption in thought may be attended by what seems to be neutral excitement, but is not really so. The intensity of thought may press into consciousness as a knowledge and feeling of intensity, but so far as it is a feeling it is indubitably pleasure or pain. This pleasure or pain may remain as continuous undertone with frequently repeated intrusion into full consciousness. Careful analysis in this case shows that apparent neutrality results from a strong attendant recognition, or from the natural volitions being quickly overruled by feelings consequent upon other considerations. 53Intellectual men are not apt to be guided by excitement. Professor Bain says that imitation is a test-case, that this is a volition which is obviously stimulated by neutral feeling. In some cases imitation seems clearly a mechanical, ideo-motor affair, an instinctive action without either conscious feeling or willing. In all other cases of imitation analysis will show excitant pleasure or pain. As Preyer and others have shown in the case of young children, mimicry arises mainly from pleasure in activity as such, and not from its peculiar quality as imitation. For children, and often for adults, imitation is simply a method of joyous and novel activity. The stimulant in higher grades of imitation is pleasure in attainment. As far as excitement is stimulant, it is, on the general principle before stated, either pleasure or pain. The pleasant feeling of intensity will tend toward continuance of imitative action, the unpleasant toward discontinuance. The pleasurable sense of activity, as inciting and continuing will in imitation, is a good example of excitement as feeling of volition-intensity.

If volitional excitement as instanced in imitation, and cognitive excitement, as exemplified in surprise and absorption of thought, cannot be termed neutral, it is quite unlikely that we shall find any neutral feeling-excitement. A person at a horse-race may at first have so small a degree of pleasurable hope and painful fear aroused that the intensity does not force itself into consciousness. The increasingly rapid pendulum-swing of consciousness from hope to fear and back again becomes soon so intense that this objective intensity of feeling forces its way into conscious life as feeling of intensity. This excitement may be mainly regarded as accompaniment, or it may be valued in itself as excitement for excitement's sake. This absorption in the feeling of intensity is eagerly sought for by the *ennuyé*. The devoted theatre-goer often induces both pleasures and pains simply for 54this resultant feeling of tension which he regards as enjoyable for its own sake. Feeling-excitement in the simpler and earlier form and in this later artificial form is plainly pleasure or pain coloured by slight element of cognition as recognition of intensity, and by volition in continuing or in stopping the causative activity.

Bearing in mind the analysis of excitement just made, the true interpretation of several matters which have been suggested is obvious and clear. Mr. Johnson (*Mind*, xiii. 82) remarks that very intense mental pleasure and pain tends to run into a state of neutral excitement. This I interpret as the mental law that intensity of any mental activity, of any pleasure or pain, tends to displace this activity by feeling of intensity. This feeling of intensity is indeed neutral as regards previous states—that is, it is not, of course, the feeling whose intensity it feels; but, as I have sought to show, it is nevertheless always pleasure or pain. Again, as to the question whether states of mind equally pleasurable or painful may have different degrees of excitement. If excitement means here subjective excitement, then I answer that they do not have any degree of excitement, for feeling of intensity can never be a quality of the feeling whose intensity is felt. If excitement is the objective form, and refers to the intensity in general, then, as has been before said, it is a confusion in terms to apply the terms pleasure and pain to it. The anticipation suggested by Mr. Johnson as a case of neutral excitement is precisely analogous to the case of excitement at a horse-race, which has been analysed. Mr. Johnson concludes that feeling is not only more or less pleasure or pain, but also more or less excitement. The proper way of stating this is: all feelings, including the feeling of excitement, consist of pleasure or pain and have degrees of intensity.

Again, let me note the relation of intensity, and consequently feeling of intensity, to quantity of consciousness—a 55subject suggested by Mr. Sully (*Mind*, xiii. 252). The fundamental properties of consciousness—quality, quantity, intensity—and also their inter-relations, would be a fruitful theme for extended discussion. I think that the clearing-up of many problems would result from thorough investigation and careful definition in these points; but at present I can only offer a remark or two upon the subject. It is plain that intensity varies with different qualities, that certain kinds of mental action are more generally characterised by high degrees of intensity than others. Presentations tend to higher intensities than representations, and pains than pleasures. It is noticeable that our psychological nomenclature, both popular and scientific, is mostly concerned with qualities, which shows that quantities and intensities have not received the attention they deserve, and have not been carefully discriminated. A representation of the same house comes up in the minds of two persons, one of whom has lived in it, the other merely seen it several times. Each psychosis is as representative as the other: they have the same quality, but in quantity and intensity they vary greatly. In a single multiplex act of consciousness, the former embraces a wide reach of detail and association and a high degree of intensity which is lacking in the meagre and faint image of the latter. Physiologically, quantity is as the mass of co-ordinate coincident activities of brain in highest centres, and intensity is as the arterial and nervous tension in the highest centres. Intensities may be equal, and quantities very unequal; as compare one greatly interested in a game of cards with a person watching a near relative at a critical moment of illness. Intensity of pleasurable hope alternating with painful fear may be equal in both cases, but in quantity the latter would tend to exceed. Very quiet natures are often characterised by largeness of quantity of consciousness. Other things being equal, intensity tends to reduce quantity and obscure quality 56of consciousness. Quantity, like intensity, may cause a reflex act of consciousness when it becomes so great as to push into consciousness as recognition and feeling of quantity; and as a feeling of largeness, elevation and mental power it is clearly distinguishable from excitement as feeling of intensity. Intensity is dependent on the force or strength by which a mental state tends to persist against other states which may be crowding in, and it is also closely connected with rapidity of mental movement; but it is primarily tension, consciousness at its highest stretch, specially as touching upon interest, an element more or less involved in all consciousness.

It would seem highly desirable, in order to keep clear the distinction between intensity and feeling-of-intensity, to restrict the term Excitement to the latter meaning, and substitute the general term Intensity for all objective excitement so-called. It is also greatly to be desired that the reflex states which arise from sudden or great changes in quality, quantity and intensity of consciousness, and which are commonly termed feelings, should receive more general attention from psychologists than heretofore. I have in this paper essayed something in this direction, but it is a very large field, and comparatively unexplored.

However, so far as the problem of feeling as indifference is concerned, enough has been said on Excitement and Intensity, and I shall now consider Neutralisation as giving neutral feeling, a method suggested by Mr. Johnson (*Mind*, xiii. 82), and developed by Miss Mason (xiii. 253). Does a feeling, neutral as regards pleasure and pain, result from the union in one consciousness of a pleasure and pain of equal intensities? Is there a composition of equal pleasure-pain forces so that resultant equals zero? Such a question implies a clear apprehension of what is meant by being in consciousness, and as to the possibility of perfect coincidence and equality in mental activities. It is plain that so far as consciousness is linear, neutralisation 57cannot occur. Where there is but one track, and but one train at a time, collision is impossible. Mental states often appear coexistent while they are really consecutive. It is doubtful whether pain from toothache and pleasure from music ever appear in absolute synchronism in consciousness, but they may alternate so rapidly sometimes as to appear synchronous to uncritical analysis. To a man drowning, a lifetime of conscious experience seems condensed into a few seconds. This means a consciousness made very sensitive and very rapid in its movement, and which acts like a camera taking pictures with a lightning-shutter. Even if a pleasure and pain did coincide, it is probable that in no case would they be exactly equal. In mental life as in organic life every product has an individuality: as every leaf differs from every other leaf, so every mental state is on completest observation *sui generis*. This is evidently a most delicate investigation, but I doubt whether it can ever be shown that two equal pleasures and pains ever appear in the same sense in consciousness at the same time. Practically equal pleasures and pains in consecutive consciousness lead to vacillation, and the secondary pain of alternation and excitement drives intelligent agents to new activity, or in stupid agents the alternation may be carried to exhaustion.

It is undoubtedly true that consciousness, in all the higher forms at least, is a complex; yet full and complete consciousness is probably of one element only, and the remaining portion of the nexus grades off into subconsciousness and unconsciousness. There is a network of coexistent states of consciousness in different degrees in mutual reaction, each striving for dominance but only one at a time reaching it. Some portions of the nexus, as Ego-tone, are quite permanent elements. The light of a large and brilliant consciousness may illumine a considerable area, but brightness most certainly diminishes in rapid ratio as the distance increases from attention, the 58single point of greatest illumination. A highly developed brain may sustain a highly complex consciousness, but it is only at the point of highest functional activity that we find the physiological basis of a full consciousness. While high grades of mental life are so complex, we do not find anywhere a mental compound. Two diverse or opposite elements never combine into a compound which is totally unlike either. Close analysis will fail to reveal any process of neutralisation or combination whereby we experience neutral states of feeling.

I have endeavoured to set forth the real nature of certain so-called neutral feelings; but at the bottom the question is, as was at first intimated, a matter of definition. Is it best to restrict the term Feeling to pleasurable and painful states of consciousness, or is it advisable for clearness and definiteness to widen the use of the term so as to include certain neutral states? From such analysis as has been made, I doubt the advisability. Appeal in such matters must always be made to analysis, and the advantage must be shown for a concrete example. The *a priori* idea or general impression that pleasure and pain is too small a basis for all feeling has no real weight. Moreover, it must always be borne in mind that psychology, like all other sciences, deals only with phenomena and not with essences, not with mind but with mental manifestations, not with feeling as mental entity having properties, being pleasurable, painful, etc., but with these qualities in and for themselves. Thus the metaphysical fallacy hidden in such common expressions as "pleasurable and painful feelings" is to be constantly guarded against. The feeling is not pleasurable or painful, but is the pleasure or the pain. The feeling has no independent being apart from the attributes which in common usage are attached to it, nor is there any general act of consciousness with which these properties are to be connected. As indicated at the beginning of this paper, this common tendency has its 59psychological basis in the bringing under the term Feeling some of the more permanent elements of consciousness—especially the Ego-sense—which stand for metaphysics as beings and entities having properties. Knowledge, Feeling, Will, are for nominalistic science simply general terms denoting the three groups of mental phenomena which seem to stand off most clearly and fundamentally from each other, and Pleasure and Pain are most clearly and fundamentally set over against Knowing and Willing. It does not seem that Professor Bain and others have made plain to us any better differentia.

If this definition of Feeling seems the best that descriptive classification can give us, it is certainly enforced by genetic considerations. The key to a really scientific classification lies in the history of mind in the individual and race. The greatest progress in psychology is not to be attained by the psychologist continually reverting to his own highly developed consciousness, but, as in all sciences, the study of the simple must be made to throw light upon the complex. Mentality like life is a body of phenomena whose forms cannot be separated by hard and fast lines into orders, genera, species; but there is a continuous development of radical factors. In the earliest forms of mind we find the most radical distinctions most clearly and simply set forth, and what Feeling is at first, it is by continuity of development the same for ever after. The earliest indications of conscious life show merest trace of apprehension of object, some organic pleasure and pain, considerable striving and effort. Mental evolution, like all evolution, is not by the elimination but by the expansion of its primal factors; and by the continuous amplification and intensification of these the highest development is reached. Pleasure and pain remain then for all consciousness as constant factors; and if the term Feeling is to indicate one element in tripartite mind, it must be held to this meaning of pleasure and pain. Pleasure and pain 60in their most complicated colourings from developed knowledge and will, and in their most subtle interactions, remain true to the primal type; and when we find a state of consciousness in which neither is a dominant factor, we had best denote it by some other term than Feeling. This evolutionary reason seems to me the strongest one for making the term Feeling signify states of pleasure or pain, and, as I have suggested (*Mind*, xi. 74-5), a genetic classification of the feelings must proceed upon this basis.

CHAPTER V

EARLY DIFFERENTIATION

A blind psychic life of pure feeling cannot long avail in the sharp struggle of existence, for to all stimulations it secures only two crude reactions, a spasmodic, defensive activity from pain, and an appropriative motion from pleasure. This perfectly subjective consciousness can serve only the earliest and crudest demands of life; but as the struggle for existence becomes fiercer, the more delicate and definite reactions, which can only come through cognition, are required. All that we can say as to the origin of knowledge in general is that it arose, or rather was achieved, like other conscious and extra-conscious functions, in answer to the pressing demands of the organism; and so far as we can see, it does not seem to be evolved from any pre-existing consciousness or any common basis of mind. It is a distinct type of consciousness, and so utterly diverse that we cannot trace any psychical continuity. However, we can remark this,—that perfect objectifying is not at once achieved, but cognition must be regarded as beginning in a very minute and obscure germ in some intense feeling state. Yet this germ does not seem to have a direct psychical connection with the pure feeling by which it is excited into existence, but it is a reaction to an opposite mode more diverse from pleasure and pain than these are from each other. Moreover, according to the law of evolution by struggle, this first cognition does not *come* to mind, but is *achieved* only in most intense will act, comparable for relative intensity to the knowledge 62originated by severest effort of a man in danger of his life listening to a barely audible sound, or watching a barely visible object on a distant horizon. The evolution point for all life is in stress and strain, and this is the law of the development of sensation at all times in psychic history.[A]

A. Cf. my remarks in *Psychological Review*, vol. ii. pp. 53 ff.

Cognition undoubtedly began as a very crude sensation, as the barest movement towards objectifying sense, as a pure sensation without any image form, any direct perception of an object. In the order of disappearance of elements from consciousness, we note that sensation maintains itself through a long series, and is the last stage before pure feeling sets in. As heat stimulus is increased, sense of heat begins at a certain point, and increases up to a certain intensity of the stimulus and to a certain intensity of its own, when it rapidly vanishes, and in the agony on the verge of unconsciousness is lost in pure pain. We note also that the cognition of object, of thing, disappears before the sensation of heat does. A person burning to death is for a time conscious of the fire, which consciousness at length is lost in intense painful sensations of heat; and this in turn, at the acme of consciousness entirely disappears, leaving only pure pain. Further, the rise to full consciousness, as well as the fall to unconsciousness, also suggests bare sensation as the original cognition. If a hot iron be applied to one in deep sleep, the order of waking consciousness—apart from any dream order—is pure pain, then sensation of heat, then awareness of hot object, and also of part heated and paining. In our ordinary consciousness it is certainly very hard to even partially isolate the various elements. Sometimes, however, a person will say, "I have such a queer pain; I do not know what it is." The psychosis thus indicated is evidently pain with a movement towards a sensation which yet is not realized. Sensation does not come though it is looked for; there is pain only, and unqualified save by the 63peculiarity of being unidentified. The sense of lack of sensation bewilders, because sensation is so constant for our psychic life; but in primitive mind there is no such feeling of queerness when sensation does not *come*, or it is not able to *attain* it. This inwrought tendency to sense all our pains and pleasures, and to feel the lack if we do not, is evidently the result of a long evolution. Sensation is thus seen to be an activity which we exercise to give definition to our pure feelings; there is something unfulfilled for us if sensation does not *come*, and we may thus go out for it and interpret the pain in sense form by a will-effort. Primitive mind, however, does not achieve its sensations as incited by this indefinite sense of lack-queerness or strangeness, but through pain at some critical moment to obtain a suitable reaction. All sensation is at first, as we even now can faintly realize, by a severe effort, and is not a spontaneous, incoming impression. Paradoxical as is the expression, "we learn to know," yet it contains a truth in that cognition is an attainment incited by the necessities of the organism. Necessity is the mother of invention, and knowledge is at first an invention which the organism hits upon to help it in the exigencies of experience. In early and even in later consciousness it is probable that the majority of pleasures and pains are so dull in intensity that they do not rouse sensation, and comparatively few incite as far as to perception. A close analysis of our own consciousness even will show many pleasures and pains, many vague states of uneasiness and discomfort, and many of organic pleasure and comfort, which lead to nothing and come to nothing for either sensation or perception. These states stand alone by themselves, and vanish with little effect on either mind or body. They constitute the outer fringe of consciousness where all mentality starts, and under sufficient pressure of life-interest develops into great fulness and complexity, or, when of comparatively little value to the organism, 64they disappear suddenly and completely. I am inclined also to think that close scrutiny will sometimes reveal for psychical life, as for the physical, certain entirely useless survivals, undifferentiated feelings of some types, and probably also some pure sensations.

I conceive then that the fundamental order of consciousness is not, as usually set forth, pure sensation with accompanying pleasure and pain, but the reverse—pure pleasure and pain with accompanying sensation; and only by a very gradual evolution indeed did pure feeling bring in sensation, which is thus always sequent and not accompaniment. We commonly inquire as to a sensation, Was it pleasurable or painful? but the true form of inquiry is, Was the pain or pleasure senseful? Did it attain to bringing in the qualifying element of a sensation, and in what form?

The qualifying of pure feeling to attain actions suitably differentiated for distinct forces must have proceeded very slowly, and have had the dimmest beginning. We cannot suppose that consciousness attained at once and easily to a manifold of sense, much less have had this brought to it, involuntarily received. The earliest forms of sensations were no doubt of those affections of the body produced by heat, pressure, and other elements which determine most vitally the existence of the organism. The first sensation indeed was undoubtedly not in any particular mode, but was a bare and undifferentiated form. It was some such indefinite and general sensation as we may sometimes detect near the vanishing-point of consciousness just before pure pain state occurs. For example, the sense of heat as such is lost at a given temperature for a given case, and there exists for a moment a vague general sensation, sensation *per se,* before mere pain absorbs all consciousness. Sensation at its very origin was not sense of any kind, sense of heat, pressure, etc., but a mere undifferentiated sense of bodily affection. The body is not, of course, apprehended as object, but there is a vague attributing 65and qualifying which marks the state as more than purely central, as being a real objectifying. Toothache, for instance, implies ache before the toothache, and this general aching is the type of early unorganized sensation. Pain is the essence of the state, and is throughout dominant, the cognition in mere aching being a very minor element. "I was awakened in the night by a toothache," is the objective description of a triple movement in consciousness, pain, ache, toothache. The earliest cognitive experiences were all of this very general type of sensation, which becomes gradually more definitely localized and qualified as distinct modes of sensation; pain-hunger, pain-heat, pain-pressure, and corresponding pleasure-sensations are differentiated. Subtract the mere pain from hunger state and from painful sensation of heat, and we have certain *quales* which are difficult to analyse, but which are cognitive in nature. Diverse bodily affections are *sensed* diversely instead of being *felt* in one mode, pure feeling.

We have far outgrown the sensation-cognition psychic stage, and speaking of psychic history in biologic terms, it belongs to the early palæozoic. We have yet to formulate the succession of psychic ages, in each of which some distinct psychic power attains dominancy, and produces minds as diverse from ours as the organisms of past ages are different from our own bodies. As already pointed out, it is an extremely difficult problem to realize by subjective method these ancient types. A mere general sensation is a very rare phenomenon in our ordinary consciousness, and even special sensations rarely occur in pure form. To realize what sensation of heat is for a simple consciousness, we must strip our minds bare of most of their furnishings, for all our sensations of heat are interpreted with reference to visual and tactual objects which must be non-existent for early consciousness. Sensation for us is a complex of sensations *plus* perceptions and other 66cognitive and emotional elements which lie beyond early mind, but which by an inevitable automorphism we interpret into early forms. This automorphism with the child is complete, and is never perfectly effaced even in the most accomplished psychologist. A life of simple feeling, or of this *plus* simple sensation is most difficult of realization; still we may have reason to believe that the psychic life of a low type consists wholly in repeated pains and pleasures occasionally rising so high that consciousness reaches to a vague general sensation, or rarely to a thrill of heat, or sense of hunger or pressure. Of course, in all cases we assume will-activity.

And we have to emphasize this again, that all sensation, like all pain, while always from objects is never of objects. The objective description here, as usual, does not give the inner state. Our automorphic tendency leads us inevitably to regard the order in which we perceive the organism to be effected by external objects to be its received order. But a little reflection always convinces us that this is in the nature of the case an erroneous procedure, that what happens within consciousness is not primarily any cognition of a world of objects, nor an apprehension of them in any form. Sensation, while objective by virtue of being cognition, is not in any way a realization of object, but is objective only toward the dynamic within the individual organism, and is not apprehension of static wholes of any kind. It is an objectifying to force, not to things, and this in the modes of physiological affection. It is not appreciation of a something, but of a somehow.

In the earliest stage of mind, as has been before noticed, all manner of material causes rouse nought more than a pure feeling mode; heat, pressure, electricity, sound, light, nutriment or its absence, if they attain to waken the function of consciousness, accomplish no more than pure feeling as bare pain and pleasure. It is, of course, natural to conceive that from the first consciousness, responds objectively 67in sensation in as many modes as organism is moved by external and internal forces; but a multiform sense origin of consciousness is not borne out by the general tendency and law of evolution, nor yet by such special indications in consciousness as we are able to observe. When a very young infant seems to reach pleasurably to warmth, if we are correct in positing consciousness at all, it is still very unlikely that there is sense of warmth, but the state is probably pure pleasure; and if there is sense of warmth, it did not give the pleasure, but the reverse. We believe likewise that it is probable that a consciousness response to nutriment is, at first, mere pleasure, and only, secondarily, organic sensation. Thus, warmth and nutriment effect, but only, at first, in the one mode of pure feeling, and secondly, pure sensation as general organic satisfaction. Lastly arises a differencing in consciousness for the different bodily changes. And the multiformity of stimulus and paucity of consciousness in modes while so very apparent in early mind is yet always found in all grades of psychic life. The responsiveness of consciousness is never perfected, and mind has a practically infinite field for the acquirement of sensation, for appreciating what has never affected consciousness, or which mind has felt or known only by some general mode. The infant, no doubt, has many pains for which it has no sensation values. These pains, perfectly pure and undifferentiated the one from the other, have had their occasion in a variety of physical changes. A native of the tropics, who on first touching ice says it burns, has at first but a single sensation for very diverse physical affections; but he soon attains an icy sensation, that ice feels not burning, but stinging cold. Men, civilized and educated, often are consciously affected by bodily changes of which they are wholly incognizant, the psychosis being not specialized according to the mode of change. In degraded states of consciousness, which come to all, there often appears obscure feeling and 68sensation, which is a practically single mode of answer to a very wide variety of physical excitation. In realizing the variety of external objects and changes the mind proceeds but slowly, each new form always at first in pure feeling. It is only as something affects feeling and interest that we ever come to know it or its manifestations in physiological change.

Sensations are, then, by no means such original and simple elements of mind as often conceived; but they are developed forms of some general undifferentiated cognitive state, sensation as bare apprehension of bodily disturbance, and this itself cannot be accounted absolutely original. The evolution into particular modes of sensation, as sense of heat, hunger, light, pressure, etc., is in the struggle for existence gradually achieved, and also therewith the evolution of special sense-organs. And we must always bear in mind that it is not the sense-organ that develops the sensation, but on the contrary the effort at sensing that produces, maintains, and improves the sense-organ. The eagle's eye has been developed by unceasing straining as incited by the necessities of existence felt in pain and pleasure. It is natural for us at our stage of development to suppose that the organs of sense *give* sensations and to explain the sensation by the physiological organ; but when we reflect that sensations *come* to us from the organ only up to the measure of the momentum from heredity, we see the insufficiency of purely physiological interpretation. Evolution to-day is on the same basis as evolution at any period, and as it always has been, it always will be, dependent upon a ceaseless *nisus*. It is only by painstaking effort—labour—that man progresses in sensibility, and this effort has always an incentive in some form of interest that is pleasure-pain basis. Thus it is that the astronomer's eye, the microscopist's eye, the artist's eye, is formed. The multiform sensibility of the tea-taster is attained by assiduous tasting, and the development in 69organ only follows *pari passu*. What is seemingly simple and original in sensation for us was, no doubt, like the very special forms of sensibility acquired by our specialist, achieved by the lower forms painfully and toilfully, and passed on to us. Our highest feats of sensation and insight may likewise for our remote descendants be intuitions, whose apparently simple nature may be asserted as the basis of philosophic systems. A genius is one who antedates the general stage of progress of his period by having as intuitions, as seemingly direct and simple knowledges and sensations, what is beyond or barely within the intensest effort of his contemporaries, though it may become common and easy for all men of later ages.

The moving factors, the active agents in the evolution of consciousness, are not, I think, sense-impressions of any kind; these are the results, rather than the incentives, of mental evolutions. Mind acquires its whole sense outfit, and receives no cognition whatever ready-made. It is hard, indeed, for us to put ourselves at the point of view of acquirement of what seem to us simple impressions of sense; but the difficulty is only of the same general nature as to understand how what seem to be direct perceptions of things in space are really indirect. The progress of psychology will, in my opinion, tend to show more and more that *givens* of all kinds are such in appearance only, and that mind in its essence is purely a feeling-effort.

The differentiation of action secured through sensation and its differentiations is evidently of the utmost importance to life, but still the objectivity secured is small. In the pure feeling stage, reaction is a very hit-and-miss affair, and in pure sensation stage it is but little better. Guided only by present sensations, the organism in the struggle for existence is blind to all objects, and, knowing not itself nor other objects, anticipatory action is entirely beyond its power. The growth of mind is to secure delicacy and precision of adjustment with largest time and space extension, 70and the achievement of objectification was a *tour de force* of the highest value. The exigencies of life-struggle lead comparatively early from cognition of mode of affection to the cognition of thing affecting. Perception arises to supplement sensation, and full objectification opens the way for intelligent activities. Thing or object is first, no doubt, apprehended tactually; but the sense of touch is, of course, acquired before cognition of thing touched. We, indeed, find it difficult to appreciate this, since in touch we constantly apprehend things as in contact with us; still, if in some very sluggish state, as deep sleep, when the varied and correlated life of sensation with perception is practically *nil*, a rough object be made to bear upon the body as a lump in the mattress, it is evident that consciousness begins as bare pain, then general uneasiness as bare general sensation, then sense of touch, and finally cognition of object by means of and through the touch sensation. The sense of thing touched follows on sense of touch. This general order may be illustrated from a squib in a comic paper of the day. A swell finding a friend sitting by an open window on a cold day asks him if he does not feel cold. He answers, "Ya-as; I guess I do. I knew theah was something the mattah with me; I suppose it must be cold." The threefold movement in this noodle's mind as evidenced by his words, is, first, feeling pain; second, a something the matter, *i.e.*, general sensing and objectifying thereupon; third, particularizing to feeling cold. He has simply gone back to primitive process. Touch or other sensation is in itself no more than an objectification of physiological change, and calls up no object whatever. In pure sensation there is no image of anything, but it is merely a peculiar modifying of pleasure-pain according to mode of physiological stimulus. A heat thrill does not include objectification to any existences, not even to the physical body of the organism sensing.

It is only by and through sense of physiological disturbance 71that awareness of object is achieved. Intense sensation stimulates to full cognition, to complete act of objectifying. This tendency of sensation is illustrated by the common saying, "hunger sharpens wit"; and certain it is that presentation of food objects is arrived at only by this stimulus. The earliest objectifying, no doubt, arose from a pain-sensation of some kind; but this primitive cognition of object was purely general, just as primitive sensation was purely general. A world of objects is not at first and at once attained, but only object barely as such, dim awareness of a mere mass. In the earliest stage every presentation is of a bare objectivity, so that one cognition differs from another in no wise as regards content. This mere thing, which is first full cognition content, is next to no-thing. When we try to conceive this thing we inevitably foist in some special sensation and perception, most generally sense of light and seeing; and the explication just made in the previous sentence was undoubtedly understood by the reader in visual terms. Our apprehension of object is correlation of several modes, and it is most difficult to intimate in any wording what bare undifferentiated apprehension of object may be. If the embryology of mind were more thoroughly studied, we should understand in some measure, for this stage most probably occurs in the very earliest activities of every human and animal mind. A *totum objectivum*, which is thing and nothing more, is, perhaps, occasionally observable in our own consciousness when at very low ebb—at such times when pure feeling and pure sensation become possible phases.

This general, undifferentiated cognition of object and all the special forms therefrom developed must always be accounted as coming about in no spontaneous way, but as attained and supported through will activity of an intense form. Perception of object is not in any true sense impressed from without, nor yet in any true sense is it a native 72faculty or power. It is not more or less freely constructed out of more or less given data. It is the necessities of life that bring mind to achieve full cognition; and this alone is the first cause of cognition, which is always in its inception cognitive effort toward objective realities, towards a world of things. These objects, among which and in close relation to which some single object, organism, must live—this is the common postulate of all biologic science, psychology included— constitute a world. The living object is such by virtue of the simplest consciousness, a feeling-will, as absolutely essential to any advantageous action. It is by this root-form, feeling-will, that cognition is ultimately accomplished, and not by virtue of any imprinting of objects upon mind as in some measure a *tabula rasa*, nor yet in any purely subjective construction of object. Object is revealed neither from without nor from within; it is achieved solely as a guide to advantageous action in the struggle for existence. Of course, the mind does not knowingly reach knowledge, does not foreknow it and its advantage in order to attain it; this is a contradiction in terms, and profects backward a highly refined teleology. All we do at present is to simply assume it as law that serviceable consciousnesses, cognition and others, are inevitably attained in the stress of existence. For the science of psychology, metaphysics apart, this is the best standpoint, and all we can now say. The confirmation of an organism's activity, cognitive and otherwise, as serviceable, is in feeling pain and pleasure, which is the original mode in which objects excite consciousness or consciousness reacts to them. It is in feeling as the starting point that cognition is determined and maintained. We cannot scientifically speak of any mental process as native, that is, mind itself is not native. By the very term original we exclude inborn. The first consciousness occurred, it was merely event, useful event; and if we further say it was acquired, we probably 73say what agrees best with biology as a whole. It is impossible at present to discuss whether or not mind may be a primitive vital function, for where life begins or ends is itself a most obscure problem; but whether it be primary or secondary, mind in no form is properly native, that is a pure given, but we simply say the function is displayed, as we speak of nutrition or reproduction. In the organism we see something which has nutritive, reproductive, motor processes, perhaps also consciousness processes; and so far as there is any problem as to the nature of consciousness as native function it belongs to a general biologic problem. As to the question as to whether cognition or what cognitions are original and simple in all mind, we have already excluded the whole field of cognition from this position.

Does the general objectification, the first stage in cognition of object, have any special function for the developed presentation forms of later consciousness? Mr. Ward, in his suggestive article in the *Encyclopædia Britannica*, seems to intimate that it has. He says (p. 50), "Actual presentation consists in this *continuum* being differentiated and every differentiation constitutes a new presentation." Mr. Ward in this connection sets forth that presentation-continuity in consciousness is determined by a presentation-continuum which is "*totum objectivum*." Presentation activity is fundamentally a differentiating of this constant element. We might compare this *continuum* to an ocean from whose surface rise waves, particular presentations, which subside again into the parent sea, which ever remains as the constant basis of all wave movements.

Now the question of *continua* is a very broad one. Do the early stages of consciousness, pure feeling, pure sensation, pure objectivity, remain as constituting the basic bulk of all higher consciousness, and is all higher consciousness but differentiation of these as well as from these, that is, is it no more than differentiating activity 74kept up on a vast series of levels and sub-levels? Or are we to regard them as regressive stages to which developed consciousness rarely returns? May we consider that there is a certain histology of mind, that certain primitive forms, like tissues in the body, constitute the inner and constant structure of mind?

The theory of *continua*, be it observed, in its fulness requires a numberless series of levels and sub-levels supporting one another, for a high form of consciousness pre-supposes an indefinite series of antecedent stages. While any highly differentiated consciousness is going on it must be an actual differentiating of the preceding stage, which is therefore coincident and pre-existent to it, and this latter in turn must have its supporting continuum, and so on down *ad infinitum*. The theory makes mind a wheel within wheel of bewildering intricacy. Yet mind in this point of view has a certain analogy with the physiological status of the higher organisms, for example, the human body is colonial, is constituted of a multitude of cells, a simple type of organisms, by whose consentaneous activity the whole body is animate.

One objection to this theory is that it confounds functioning with differentiating. Not every act of consciousness is by its very nature a differentiating, a movement toward specialization. Consciousness is on the whole more often regressive than progressive, and very often practically neither, as for example, in all instinctive, habitual, and spontaneous activities.

But again, while differentiating act certainly pre-supposes the undifferentiated, does it require coincidence? For instance, vision as ordinary form, receiving impressions, certainly contains no *totum objectivum* activity, but also as differentiating act, as intense visual effort reaching to higher development, it generally, at least, seems free from any lower stage, and is engrossed in itself. Since we make the prime cause of all mental development and 75differentiation in will, we do not need any undifferentiated general ground remaining in consciousness as basic element, nor does analysis of consciousness show this constant element. Successive phases of presentation development are attained through effort, but one does not gradually grow and branch out of the other by a purely inward *impetus* of its own. I believe, indeed, that the inner life of mind consists in its original forms, and that they remain in late mind not merely as useless survivals but having a distinct functional value; but I do not see how any or all of the general stages of mentality constitute *continua* for consciousness of higher types. Instead of being constant basal elements they occur and are blotted out with such rapidity that reflection can very rarely identify them (*vide* p. 63). They are lost and swallowed up in complex consciousness so quickly as to leave no trace upon memory, and they do not subsist or continue throughout the complex forms. They are then the very opposite of *continua*, being, in fact, the most evanescent of mental phenomena. Consciousness in all higher forms, as the human mind, must and does mount the main steps of its very early growth with marvellous rapidity and leaves them entirely behind. The more primitive the stage the more quickly it vanishes, till often it seems to appear in tendency form only, or be thrown into a subconsciousness. Primitive types exercise a most important but fleeting influence in advanced consciousness which rises through them most rapidly and easily, but in the less advanced the contrary is the case. The Australian savages, as observed by Lumholtz, came to their senses and reached a full awakening in the morning very slowly as compared with civilized men. With dull children likewise we observe how slowly they awaken. All regressive forms reach but slowly to their full consciousness and dwell long in intermediate stages. But in all cases when higher forms enter the lower disappears, when varied perception enters in awakening, 76then the preceding dim general objectivity is wholly obliterated.

It will be remarked that admitting, as we do, the constant existence in mental life of feeling as pleasure and pain, we thereby make this a real *continuum*. But we may say that consciousness is never without a pleasure-pain constituent and yet not assert a *continuum*. Consciousness continually possesses some pleasure-pain element, but this is not a feeling as continuous state, as an underlying differentiating basis pleasures and pains as diverse independent states are essential incentives in all consciousness, but they do not constitute a single continuum.

Of course, every consciousness, as long as it continues, is in this very general sense a *continuum*, but no form of consciousness, primitive or advanced, can, with one exception, be called a *continuum*, as a single mode running through and unifying a long stretch of varied consciousnesses. This exception is the complex element of ego-tone. Early mind is no more than a kaleidoscopic jumble, with no one organizing and unifying element. Even when consciousness from happening in purely disconnected flashes attains first a certain limited continuity, this is not by means of some conscious element persisting through a series, but merely signifies that as fast as one consciousness dies out, another takes its place, *i.e.*, the continuity is purely formal and temporal. It is through self-consciousness alone that any real *continuum* is achieved in and for consciousness, and this ego-tone is far from being primitive.

The sensation and objectifying as discussed in this chapter in connection with feeling, both pain and pleasure, constitutes complex states of consciousness which may be termed a feeling when the pain or pleasure is dominant, or a cognition when the sensing and objectifying is dominant. Thus by a feeling I understand a state of consciousness which is either entirely or dominantly pain or pleasure, 77the former being pure feeling, the latter mixed feeling. This latter class constitutes the feelings properly so-called, as varied pains and pleasures, the variation element being the cognition in some form. Feeling as being in different kinds is made such by the differentiation of cognition. Thus hunger is neither a pure sensation— that is by pure sensation meaning not absolutely pure, for pleasure or pain is invariable incentive concomitant, but sensation pure from any distinct mode of apprehension, as merely general and undifferentiated—nor yet is hunger pure pain, but it is the combination of a certain definite sensing, beyond the pure stage, with pain. Hunger is a feeling when the pain aspect is dominant, is cognition when sensation aspect is dominant. The confusion in the use of the terms sensation and feeling comes from the difficulty in determining dominancy in given cases. Certainly the exact line where feeling of hunger passes into sensation of hunger can be settled only by the most careful discrimination, but at any great remove from this line the character of the state is very manifest. By no effort can we separate the sensing from the pain so as to have nothing but sensation, though the attributing to bodily affection does in the incipient stages of hunger become dominant, but as hunger increases, pain becomes dominant, and ultimately the end as the beginning is pure pain. We say, "I *feel* hungry," for all stages when any sensing is present, and this indiscriminate popular use of the word "feel," has tended to obscure the real nature of the whole mentality. The same line of remark applies to feeling thirsty, feeling hot, etc.

CHAPTER VI

REPRESENTATION AND EMOTION

"I feel cold," and "I feel afraid of cold," are expressions which denote two tolerably distinct feelings. The main characteristic which distinguishes the second feeling as an emotion is obviously representation. In the first case, I have pain with presentation of the cold, in the second, pain with the mere representation of the cold. If I feel cold, I have direct and immediate experience; if I fear the cold, I have an experience in view of experience, pain at pain. When one says, "I have a violent pain in my head," and a friend answers, "I am deeply pained to hear it," we recognise at once the fundamental distinction between sensation and emotion. We have in this chapter to discuss some points as to the rise and nature of emotion in its relation to representation.

The theory which we have been elaborating is that pure pleasure and pain are the original and causative elements in the whole realm of mind. Pure feeling is the most direct and necessary, and so the first response in conscious form, to all stimuli, and it is the incitement to all cognitive activity in its inception and growth. The harm and good to organism, are at once, and most quickly realized in terms of pure feeling, and the painful necessities in the struggle for existence, lead to a continuous development from this point. Dominant pleasure and pain, with the different presentation forms, constitute different feelings, as of warmth, hunger, cold, etc., to which some fuller objectification may be added. Adjustment is thereby 79made manifold, but only with present stimulus. There is no appreciation of the experienceable. All that is attained is immediate present apprehension which in no wise suggests or interprets, but which is strictly self-contained.

We must, indeed, acknowledge that no consciousness, save, of course, the very first, can exist in perfect isolation totally unaffected by any other. The second conscious activity was not a perfect facsimile of the first, and its variation is due at least in the main to the precedent mentality. What is, is determined by what has been, and this universal law is in mind the inductive nature of all experience. The solidarity of all mentality and of all materiality is a scientific postulate, a principle which we must assume, or deny all scientific investigation. The movement of a molecule in the sun, millions of years since, influences the condition of my body to-day, and the flush of pain in some protozoan millions of years since, has had an infinitesimal share in determining my present state of mind. Yet this fact that every psychosis is what it is by reason of the whole line of previous psychoses, does not lead us to suppose that experience cognizes itself from the beginning, and consciously builds itself up. There is for a long time no consciousness of process of mental integration. The whole universe of mind is the necessary *prius* of each individual manifestation, yet the particular phenomenon in consciousness does not include a sense of, or reaching out to, these conditioning agencies. No sense of dependence is generated. But we ask, How can one conscious state unconsciously effect or determine another? How can consciousness be affected without consciousness of affection? Yet, difficult as it may appear to set clearly before us the nature of this relation of a consciousness to all the preconsciousness, it is still obvious that the intricate *nexus* of cause and effect in mind does not need to be known 80of mind or realized in the individual consciousness, and is not, and cannot be. Every consciousness is the derivative resultant of innumerable pre-consciousnesses, and it goes to the determining and qualifying of innumerable post-consciousnesses, yet it is neither consciousness of the future or the past, though it involves both.

The early phase of mind where consciousnesses are wholly un-unified from within by any central or continuous consciousness, and whose solidarity is wholly in an unconscious integration is so foreign to us who have minds where experience of experience is continually in process, that it is with the utmost difficulty we can in any wise conceive it. It is evident that a very low organism may have consciousnesses, but no mind, that is, no self-unifying whole of consciousnesses. It does not possess a mind, but during its whole life it attains psychoses which are merely *disjecta* reached to help an immediate necessity of existence, and then fading completely away. Each psychosis is achieved more easily than the former by reason of the former, though there is no consciousness of connection with it. The increment and qualifying of a given experience by past experience is not reached by it. Some differentiation is attained under pressure of struggle for existence, and experience is constituted, but is wholly unknowing of itself and in no wise self-formative.

We have now, however, to consider the problem, how experience came to itself, and how and why representation and emotion should arise in the struggle of existence.

At the first, as we have seen, organisms responded in conscious form only in pleasure and pain, and this only when the actual damage or benefit to the individual was very considerable. When the hurt was critical, then only was pain accomplished as a function to secure self-preservative action, but gradually through survival of the fittest the greater susceptibility was attained, so that minor lesions are felt in pain terms, and some general sensing 81and objectifying lead to some differentiation in adjustment. The external parts of the body become specially sensitive, and ciliate extensions are formed. Injury to these results in pain and consequent reactions, and in this wise by injury to a small part great harm to the organism as a whole is prevented. The low forms of life are thus enabled to avoid the hurtful before they meet it in full annihilatory force. These practically anticipatory reactions—though there is no real anticipation in consciousness, no real experience of experience—I term the method of incipiency. Pain reactions are thus reached with less and less actual harm until the very slightest injury to a minute tentacle will suffice to awaken pain.

This tentacular experience, however, is obviously very limited, and has incidental disadvantages. Further, that pain should be attained when there is little actual harm, is good, but to attain pain, and self-conservative action before any injury is done, but only about to be done is better. Reaction to potential harm is a most important advantageous step. In the earlier form of mentality, the animal must actually be in the process of being devoured by an enemy before a pain reaction is achieved, but in the later representative form of reaction there is complete anticipation, and the animal can come off with an absolutely whole skin. Ideal pains, as fear, anger, and other emotions, are gradually substituted for pains which are real in the sense that they arise in a positive hurt to the life of the organism. The saving which is effected through emotion is most important, and this economy is reason for the rise of emotion in the struggle of existence. Those animals who are able, not merely to react on slight injuries to themselves, but also through fear, etc., to avoid all actual injury, have a very manifest advantage.

If now the *rationale* of the rise of emotion is apparent, let us next proceed to some analysis of emotional process in general. The mental mechanism by which anticipatory 82function is secured is certainly complex, and a complete analysis presents many difficulties.

In the incipiency stage, which we have just discussed, the organism was enabled to avoid the full force of the injurious by meeting it half-way with extensions from its own body, but we cannot suppose that this was purposely accomplished, or that the lesser pain conveyed in any form sense of the greater pain. There was no fear, no anger, not any experience at experience in consciousness. There is simply pain on less and less injury, but no anticipation of pain.

In early consciousness there is, of course, frequent return of a given object which becomes the occasion of a large number of objectifyings which are identical in nature yet do not contain sense of identity. There is repeated reaction to the same objective stimulus, yet with no sense of sameness, there is frequent cognition of the same thing yet no recognition. With primitive consciousness, no matter how often a thing is experienced, it is equally new; revival of the past is not stimulated, nor sense of identity attained. Mere return of a state is not sense of return, and no amount of re-occurrence or combinations thereof will make sense of re-occurrence. Re-occurrence of a psychosis is nothing more subjectively than occurrence unless there arise sense of re-occurrence or revival. The pure feeling states in primitive consciousness are perfectly identical in nature, and they arise on occasions which are the same, yet there is of course no sense of identity. A young child may see a thing a hundred times without recognising it; there are a hundred re-occurrences of state yet no sense of re-occurrence. The hundredth perception does not differ materially from the first, does not include any true representative element. The immediate image does not stand for the past, the mind does not revive previous presentation on the strength of it.

83Mind is regarded by many as consisting fundamentally of vivid sense presentations and their faint reproductions, of sense impressions and their representations. That which has been repeatedly experienced has a tendency to re-occur without the particular objective stimulus, but merely indirectly by some connected stimulus, through an association of states. But this revival, however attained, does not constitute real representation, it does not really differ from the presentation simply because it re-occurs without the original particular objective stimulus. Representation in true sense of term is representation with sense of re-presentation. A representation is a repetition of a presentation with no consciousness of repetition or any added nature. Repetition is a fact in consciousness before it is a fact for consciousness. All presentations and re-presentations have mere immediate validity and value, they point to nothing, and mean nothing, there is no going beyond what is immediately given, no prescience of a possible experience.

Revival often occurs in mind without sense of revival, and so is not true representation. In disordered states of the nerves we frequently see objects which have no real existence, the states are revival states as objectively interpreted, yet there being no sense of revival they stand in consciousness for real presentations. When I see a person sitting in a chair but afterwards find that no one was there, I characterize the state very naturally as a mere imagination, a representation; yet in fact it was in subjective quality a presentation. We are not to psychologically classify, as is too often done, psychical states according to presence or absence of object, but as to sense of presence or absence of object. It is only as consciousness takes note with reference to object that there is differentiation in consciousness to make presentation and representation.

We must consider it probable that the earliest revivals 84by consciousness were solely of the unconscious sort, or, objectively speaking, were hallucinatory. A sense order is formed, which attends to a series of objective realities; let now, on some occasion, one of these objects drop out, yet there will be attaining of some sense of it as though it were present, and the proper reaction will be carried out. The mind gets its early revivals without sense of revival. They have presentative force, and are sensings of objective reality though there is no objective reality there at the time to sense.

These early simple revivals, which are all hallucinatory, perform an important function. They are practically anticipatory, in that the reaction is secured before the actual presence of the reality. Thus they save an actual bodily experience, though the mental is quite real, yet fainter than actual object would give. Thus with an enemy an animal will revive, upon slight indirect sensation, previous experiences, and it will have in ideal form, *i.e.*, without the objective reality, a very real experience with what is to it real enemy, thus escaping before full advent of enemy. When a shadow alarms a low organism—and even very low organisms seem to react to shadows—there is no actual harm done to its members as would happen with a concrete body, and hence there is no direct pain. The shadow is yet taken for real body, and revival pains and revival sensations are attained with this, and there is consequent activity. Shadow does not appear as sign of enemy, but in itself a dangerous reality, so that anticipatory reaction is gained without actual representation. In most cases in low organisms what we take for fear or other emotion is probably no more than revival of the type of which this shadow experience is an example. What is actually unreal, being only revival, is taken for the real, and is acted on accordingly, and in most cases this action is of service as anticipatory. When the organism discovers the shadow to be but shadow, a something, 85not the object, yet connected with it, when it becomes a sign of further experience, this is representation as the basis of emotions such as fear and anger.

The pain intensity in the simple revivals, re-presentations, is doubtless less than in experience with objective realities, so there is a saving on this score in pseudo-direct experience. While reactions are secured upon this method without injury being actually inflicted, still there is loss of economy in this, that the activity is excessive under the circumstances. Priority of action to real injury is secured, but at an excessive expense of energy, almost equal to that in actual experience with the real thing.

This acting to a false reality, while it has a value for experience, is, as said, uneconomical, and it must sometimes not have the anticipatory force. The cheat and illusion is ultimately at some critical moment cognized by consciousness, revival comes to be estimated at its real worth, and sense of reality and unreality is formed. The revived presentation does not stand in and by itself alone, but it acquires a significance, and it loses the force of complete reality value. That which is brought into consciousness again is not only revival, but is felt to be such.

To constitute representation, then, there must be not merely revival, but sense of revival with some sense of unreality of revival form. But this would avail nothing save it brought in sense of its value for experience. The revival must not only be appreciated as such, but the relation to the experienceable must be cognized. The calling up of the past must be applied to experience. The sight of a fire not only calls up revivals, but there is the sense of the experienceable therewith, and an emotion which incites me to walk to the fire and receive warmth. Mere return and sense of return must be supplemented by sense of value for future experience. Representation is experience doubling on itself. All representation is more than representation of thing, revival; it is representation 86of experience as such, hence an experience of experience. We must always emphasize as the essence of representation not the revival, but the sense of the experienceable or experienced thereby conveyed.

The process to representation we see exemplified in measure in awaking from a dream. The dream itself, speaking from the objective point of view of observing psychologist who detects no real things in interaction with the body, is representative in nature; but, for the experiencing consciousness, there is no sense of revival, and all is presentative activity. Things are known as such, and not as dreamt or represented. Awaking is a gradual pouring in of sense of revival and of sense of objective unreality of the experience; we become conscious that the activity is no direct consciousness, but a recalling or reproduction. The dream image, which was so real to me while in the dream, I now hold as representative only, as having no immediate answering form and substance. When, as with the superstitious, the dream is felt to have significance, to have a meaning for life in pleasure-pain terms, then emotion becomes possible, and fear, hope and kindred feelings are excited.

We observe that representation is then a new order of consciousness. Representation cannot be attained by any combination of experiences, revival or direct, but it is a unique and reflex act. It is not a development of presentation, as an echo and re-echo of it; and the mere fact of absence of external cause or object does not constitute a cognition as representation. The objectifying is not self-contained, but it conveys a meaning for experience. Representation is an experience which includes some cognizance of or sense of experience, and it is thus the germ of self-consciousness and consciousness of consciousness. Experience comes to be more than a series of detached and isolated activities with no cognitive power beyond a direct and immediate apprehension, but by rising to some 87appreciation of itself it becomes forewarned and forearmed, able to consciously appreciate and attend to its own welfare.

We have also to emphasize this, that while representation involves a conscious re-objectifying, it must also include some re-feeling consciously accomplished of pain and pleasure. Revivals of pain and pleasure are felt and are appreciated as revivals, as having their basis not in present object, but in previous experience. It is by understanding feeling as experienced and experienceable, it is in view of pleasure-pain experience, that emotion arises. It is not sense of imminence of object, but of imminence of pain and pleasure, that awakens responsive emotion and so self-conservative action. Emotion always implies a pleasure or a pain in ideal sense of the experienceability of either. Representation as cognitive revival and sense thereof is subsidiary to representation as feeling revival with sense thereof. For instance, the representation of a tooth and of pain of toothache are correlative representations. Mere representation of cognition has no value in itself, is a mere idle panorama, save as it brings on representation of pleasure-pain. Unless representation of object implies representation of pain, there is no deterrent effect on the mind, and no proper bodily reaction.

We may believe that the order and basis of the representative side of mind is practically the same as indirect and simple activity, that the actual motive forces and originating impulses are pleasures and pains. We should suspect that the first revival attained was a pure feeling revival, and that the first representation was of pain and pleasure, and not of object, a consciously re-feeling rather than a consciously re-objectifying. The immediate value of the feeling side necessitates that all differentiation be initiated there.

Representation is only of experience of things or of pleasure-pain experience. It is always experience of 88experience, hence the expression, representation of an object, is, in strictness, inaccurate. Experience of things, as cognitive act, is always presentation. Yet early representation must be considered as very much adulterated by presentative elements. It was only slowly that representation was differentiated as a distinct power such as we find it in human consciousness; at the first it must have resembled the confused state that we sometimes experience between sleeping and waking when a given image often shifts from presentation value to representation value, and then back again.

Representation at the first is also purely concrete and particular. Bare appreciation of the experienceable does not include idea of experience. But representation in itself is merely a calling up and application of definite experiences as such. Experience as general term is not known, but only the particular facts as experiences.

The earliest emotions arise, of course, with reference to the bodily functions which have the most direct vital significance, as nutritive, reproductive, and motor activity. Very simple organisms seem to apprehend that a certain object is food before actually consuming, to have sense of the experience, and some emotive disturbance. The pleasure of feeding and incorporating into the bodily tissue is sensational, but any feeling previous or subsequent to this and with reference to this is emotional. A very young child feeds, and does not know food. Gradually it associates the visual sensation of whiteness of the milk with the immediate taste sensation and pleasure feeling. But the sense of whiteness at first arises only with and after the actual taste and pleasure experiences; it only gradually notices what gives it satisfaction or pain, thus repeating the evolution of mind, which is from feeling to sense, and not *vice versâ*. Only slowly does it attain power of appreciating whiteness previous to actual experience and as *indicative of such*, that is, a power of representation. 89Then emotions, as expectancy, and desire, become possible, and will can be stirred to active appropriation of food, a fact of the greatest importance in the struggle for existence. Once attaining the sense of the representative value of its cognitions, the child is enabled to consciously accomplish anticipatory actions.

An element which complicates emotion at a late stage is representation of representation in indefinite *regressus*. In advanced human consciousness, where mind is very reflective and introspective, this phase is prominent. The *nuances* of modern emotion are largely due to this mode of complication. Montaigne remarks that what he most fears is fear. As fear implies representation, fear of fear implies representation of representation, which in its turn may be feared, and so on *ad infinitum*. Spencer terms love of property a re-representative feeling; but this psychosis does not imply representation of representation, but merely representation of desirable realities. Desire of possession is an emotion, but not emotion at emotion. It is not an experience in view of representative experience, but with reference to a direct experience, that of ownership. Since we make representation the basis of emotion, it would be natural to make classes of emotion representative, re-representative, etc.; but this is quite too subtle a distinction to be fruitful or practical.

As there are stages of representation, so there are varying degrees of strength in the sense of representativeness. A colour may be recalled to consciousness several times as neither more nor less red, and precisely of the same quantity, yet the sense of its representation quality may differ greatly at each time. There are all degrees of intensity in this sense, from dimmest feeling, when the representation hovers on the confines of the presentation field, to the point of perfect conviction of representative nature. When consciousness is not exactly sure whether an object is directly seen or only recalled, is a presentation 90or only a revival, sense of representation is obviously at its lowest degree of intensity.

We have also to remark that in presentation and representation the object is not to be divorced from activity. It is a natural analogy that cognition as subjective-objective is a picturing, the picture and the object pictured seeming to be diverse but co-existent constituents of consciousness. Cognition seems to consist in both the thing as realized and the realizing act. It is an attitude of mind which is a holding on to a something which it has in its grasp. But there is no distinction in consciousness itself of the presented and the presenting, the represented and the representing, of product and process, of content and activity; there is only the presenting, the activity, which is itself the object. Sense of colour conveys, indeed, by the common vice of language that the colour exists for consciousness, and is perceived by consciousness. But, subjectively and psychologically speaking, the object is always no more than the objectifying, the thing no more than the activity. Thus the analysis into content and activity is fundamentally false; it assumes a world of objects which are merely at bottom object-sensings.

Emotion is an arousing and energizing. It is perturbation, disturbance, agitation, excitement. It is a throwing open the throttle and putting on a full head of steam. The whole organism quivers with the sudden inflow of force and life, is quickened to its highest pressure. In all higher psychic life it is a driving force of the utmost importance. However, the trend of evolution is in the direction of economy, and with the highest forms of consciousness emotion accomplishes its work even before arriving at agitation intensity. Feeling of the emotion type, that is, representative, is always at first a rather intense perturbation. Fear, for example, is with the lower minds always fright; with higher minds it often appears as dread. I stand on the railway track when a train is 91approaching, and a slight fear enables me to take the self-conservative action of stepping from the track; but with my dog, in similar circumstances, I judge by his hasty jump and general expression that his fear is always more intense and more generally disturbing. Emotion being a force which quickly tends to exhaustion, it is obvious that those animals will, *ceteris paribus*, have the advantage which react with the least expenditure. Thus the tendency of evolution is away from intense emotionalism.

In this emotion conforms to a general law. The earliest occurrences of any given form of psychosis are with strenuousness and with exaltation and excitement of the organism. We speak of fits of anger and gusts of passion, but for early consciousness we might also justly speak of fits of seeing and hearing. Common vision of external objects is for lower consciousness as rarely attained, and requires as much of force as beatific vision of seer and poet in the human mind. The new psychosis is but momentary, and implies high tension and great friction, but progress is toward continuity and ease of working. Emotion is in human life a tolerably constant element, like perception with whose representative side it is correlated, and within certain ranges it rises because of the force of heredity with apparent spontaneity.

We remark that the social significance of emotion is embodied in the word *treat*, as *treat* kindly, badly, etc. Our *treatment* of each other always means activities inspired by some emotion.

We must acknowledge that representation is very complex and difficult of analysis. For our present purpose, however, representation is a revival with sense of revival and unreality, and yet indicative of reality experienceable in pleasure-pain terms, and thus the occasion of emotion as stimulus of self-conservative action. The young child perceives no danger; its pleasures and pains are not related to things, and have not led to the evolution 92of a world of objects. Pain and pleasure lead it slowly to correlate its senses, so that the burnt child learns to dread the fire; the emotion of fear is aroused with cognition of the experienceable. Objectively, we must divide psychoses into those which directly result from actual engagement of the organism with objects, or the reverberations therefrom; subjectively, into simple self-contained states, and into reflex states which view experience, and so being representations involving emotion. Just how from re-experience sense of re-experience and of its value for experience—sense of pre-experience—arises, is something we have not particularly inquired into, but it is something that appears a mysterious and difficult problem. That the perception of object should ever carry with it sense of possibility or certainty of further experience, painful or pleasurable, is, when candidly considered, a remarkable and singular operation. The problems of origin of consciousness of self, of consciousness of consciousness, and of sense of reality seem unsolved, but I believe that a thorough study of representation would throw much light on these points; but this is not the place to pursue this investigation. When we take up representation—emotion life in detail, we may be able to make suggestions on some moot points.

CHAPTER VII

FEAR AS PRIMITIVE EMOTION[B]

It may be considered as plausible that if the first feeling was pain, the first emotion was also of the pain character. The first representation of an object as painful induced that reaction of mind which we term an emotion, and the painful emotion we call fear. That the first emotion to appear was fear, as fright, seems likely when we consider that the general alertness and defensiveness imperatively required in the struggle for existence is thereby most immediately and simply attained. The acquirement of the power to become frightened is plainly a most important requisite for self-preservation, and thus is indicated as a very early factor in conscious life. An animal being devoured by another may merely suffer pain without any perception of the object as pain-giving and to give pain; but if it attains this perception, there may be added to the stimulus of simple pain that of fright. The direct actual pain may be but small, and so inducing but feeble reaction, as when some less sensitive portion is being injured; but if there occurs a vivid representation of potential pain, fright happens and stimulates most strenuous endeavours, and so rids the animal both of the immediately and the prospectively painful. Thus emotion acts as a complement to simple feeling, and also secures practically anticipatory reaction. Animals which must receive actual injury before experiencing pain are clearly 94inferior to those which experience emotion-pain before the injury is actually received. Other things being equal, the most easily frightened have, in the midst of many destructive agents, the best chance of survival and of perpetuating their kind.

B. Originally appeared in part in *Philosophical Review*, i. pp. 241-256.

It is unnecessary to dwell at length on child life and savage life as illustrating the primitive quality and function of fear. The earliest experiences of the child with things are lessons of fear. The burnt child dreads the fire, and thus is enabled to preserve himself from threatened injury. Fear is a primary and most important motive to action in a very wide range of the lower mental life. Those who have observed animals and man in a state of nature are always greatly impressed with the constant and large part which this emotion plays in their consciousness. With the timid and weaker species, like the rabbit and squirrel, it is likely that a majority of their cognitions prompt to fear or are prompted by fear, and with some persecuted races of savages the same may be said.

The necessity and value of anticipatory reaction being acknowledged in the struggle of existence, we plainly see a primitive motive thereto in fear, and the earliest emotional life which we can clearly interpret likewise seems to be fear.

It is sufficiently easy to see the general function of fear and its primitive character, but we find it very hard to make a satisfactory analysis, and to show the exact steps of its evolution. It is obvious, however, in the first place, that fear, like other emotions, is purely indirect and secondary experience; it pre-supposes previous painful experience of the feared object. Pain experienced in connection with cognition of object is the basis of all fear. Animals that have not felt pain from man do not fear him. But fear while thus based on previous direct experience is always hindered by simultaneous direct experience, as, for 95example, sensation. Thus when we, whip in hand, say to a child crying from fear, "I will give you something to cry for," we imply the law that direct pain and sensation tend to supplant indirect feeling as emotion. This common expression emphasizes the essential representativeness of emotion, its imaginary nature, as also the supplanting power of direct real experience. The sight of the whip inspires fear in the child who has been whipped, but this fear is in the course of a punishment wholly eliminated by the direct pain endured. The direct experience is thus the basis of every fear, but only as it is cognized, and not felt.

The great difficulty in analysing fear is in clearly apprehending the mode in which previous experience is utilized. If we could study in ourselves the genesis of a simple emotion, we should doubtless be enabled to see the steps by which experience reacts upon itself so as to give a reflex form like the emotion of fear, but this is hardly possible. However, cognition is evolved at the instance of pain, and all objects are viewed, not for themselves, but in their feeling significance. Cognition is embedded in feeling, and at first is a mere tone of feeling. Things are not at first known for themselves but solely as sources of present pleasure and pain. Things are perceived in and through the feeling which has stimulated the perception. The immediate feeling value of the object is given by the very origin and process of cognition. When an animal is pained by contact with a sharp rock, and this pain stimulates cognition of the rock, this is solely on the pain account. Repeated experiences enable the percept to arise at stimulus of less and less pain, and so the proper reaction is accomplished more and more economically.

We may say that the order of evolution is this: first, a pain; second, a cognition of pain-giver—"it hurts"—third, emotion about pain-giver, as fear thereof—"I am afraid of it." Primitive and normal cognition always 96implies emotion as impelling self-preservative action. Knowledge which does not spring into emotion and action is abortive. At first the known is always startling.

The original pain-impelled cognition brings in the painful emotion, primitive fear. And as knowledge has brought in fear, so fear reacts on knowledge, and fearfulness incites to knowing even when the pain from object ceases. Thus before any actual experience of an object it may be known and felt about. Thus that habit of objectivity is formed, of alertness, of a fearful sensing and perceiving, which is noticeable in many animals. A cognitive-emotive, emotive-cognitive life is formed and developed. It is a tremendous stride onward to be able through fearful cognition to wholly pre-perceive and anticipate the injurious, instead of having to suffer it in part before being enabled to get away.

Now primitive fear and all primitive emotion plainly utilizes the past experience as interpreting the future; emotion is about a known potency. Yet it is often stated that emotion is but a summation of revivals of past experience. Having often been burnt by fires that I have coincidently been looking at, it sometimes happens that I see a fire which has not yet harmed me, but still the mere sight affects me with what I call the emotion of fear, which, in closest analysis, means merely the revival of the burning pains associated with this seeing in the past. "I am afraid" equals "I re-experience the pains of burning" by suggestion. Pains faintly re-occurring constitute the painful fear. There is in this mass of re-awakenings no real cognition of experience and no feeling about it as such, no psychosis *at* the experienceable. And it is certainly true that when a fixed sequence of experiences tend to recur together, there will follow upon the cognition, revival waves of pain before any actual increase of pain is really inflicted in the given case. These waves stand for, and are the echoes of, the former 97real pain sequences of cognition. Thus the perception of a great mass of ice will often cause a shivery feeling, a painful sensation is revived as correlated with former cognition experiences. Even the image or representation, the purely and consciously ideal cognition, may bring in painful feeling, as when I say, "It makes me shiver to think of it." Here the painful sensation-bringing idea is cognized as such, but the representation here is the occasion of a direct painful sensation, and evidently does not imply fear or other emotion.

While not arising from actual injuries, revivals strengthen both cognition and volition. They have recurred before further hurtful experiences with the fire which originally incited them. These revival pains of previous sequences to the cognition, which are carried along with the present cognition, are real enough in themselves, yet they are objectively anticipatory of actual injury. The whole order of previous experience is by the nature of mind and nervous system re-enacted before the actual injuries are inflicted. It is always a race between mind and nature, but it is a prime function of mind to anticipate practically the movements of nature. Mind by its revival forms accomplishes this, but if it lags in its work the real injuries are mercilessly inflicted by slow but sure nature. When the sequence of revival is quicker than the objective sequence, the reactions anticipate objective order, and thus a manifest economy is achieved. But pain revivals of this kind are not fear, nor is there a real pre-perception. Since the revival forms are, to the observer's point of view, incentive to anticipatory reaction, psychologists must often, especially with low organisms, mistake them for fear; the animal is often, doubtless, merely suffering revival pains when it appears to be fearing pain. Thus we may suspect that organisms which seem to fear shadows or real objects are often merely suffering revival pains brought up in conjunction with the cognition, and not 98really fearing as result of perceiving feeling quality inherent in the object. Manifestation of pain must often be mistaken for manifestations of emotion, and there is as yet no accurate objective determination for fear or other emotions.

Revival pains are not representations of pains as in some way coming from object. Emotion requires representation, and cannot occur in any presentation or re-presentation chain. True pre-perception is not merely perceiving the thing before its effects in feeling are experienced, but it is a *representing* the feeling quality of the object before, in any given case, this quality is directly experienced. This obviously rests on past experience, but the connecting of object with pleasure-pain experience is at all times, as before intimated, equally a problem. Emotion and representation are built not of revivals, but upon them perceived as such. At some critical moment, in some rather early period in mental development, a consciousness which was pain *plus* sense of object, realized, under the pressure of struggle for existence, the feeling quality of the object, and there arose with the knowledge of object as pain-giver the painful emotion. And as soon as object is not merely cognized, but cognized as pain-giver, it may be feared. The moment that object was known as a pain agent, then fear of the object came, and thus true anticipatory action arose. We are said, indeed, to fear objects, to fear men, animals, etc., but, in truth, the fear is never of the object as such, but only in view of its pain agency. The cognizing the experienced and experienceable as such seems then a peculiar and distinct process in fear and in all emotion, a *genus* apart which cannot be constituted by interaction of simple elements. The growth of mind is largely in multiplying and enlarging the signs of experience.

The connecting once achieved of object with pain, it becomes increasingly easy to cognize the feeling value of 99objects, and before full and extreme pain experience therefrom to pre-react through emotion. Thus emotion saves both direct pain and injury. As it becomes a permanent tendency, and an impulse of consciousness to proceed from all pure feelings to cognition of object, so also to cognition of object in its feeling quality, and thus by inherent tendency it ultimately comes about that there is attaching of pain to various objects cognized, even when there is no immediate experience of pain to be connected therewith. Finally the precedent inciting pains to cognition become such minor factors, and knowledge arises with such apparent spontaneity, that emotion as involving pain significance becomes dominant rather than the immediate pain. An order of consciousness becomes established in which the notable event is emotional cognition of experience values as bringing in permanent emotion rather than an order of pleasure-pain inciting cognition with evanescent emotion. But at the first it is evident that fear was but a slight event in a consciousness which was mainly absorbed in immediate pain experience and some sense of object. It is so habitual and instinctive for us to perceive all things as having feeling value, that it is most difficult to appreciate the standpoint of a consciousness which is just attaining emotion life.

The preliminary elements to simple primitive fear, as expressed by any such phrase as, "it hurts," are at least four: pain, cognition of object, cognition of the pain, cognition of the pain agency of object. These operations, as being at first successive, do not necessarily imply, however, sense of time. The consciousness of a pain is certainly, at first, consciousness of pain really past, yet not consciousness of it as past. The pain stands as immediately antecedent act to the consciousness which is cognition of it, but sense of experience is not thereby sense of experience in time. The sense of time-relations of experiences is wholly subsequent to the simple sense of experience. All 100experience is, of course, in time, but far from being of time.

An organism, which has suffered knowingly from an object, and so feared, attains at length the power of fearing antecedent to any real injury. This seems to be brought about somewhat in the following manner: If I in any way, as by a pin pricking, rouse a sleeping animal to a cognition of an object which has often injured it, and which it has often feared, immediately there would re-occur the original concomitants of the cognition in the previous cases; there would be pain, cognition of pain, ascription to object, and fear, all merely revivals, and happening most probably before any actual injury, etc., received in the present case. Now these revivals, as before insisted, do not and cannot in themselves alone form a new fear. This is only constituted when the revival pains are known as such, when they are not merely presented in consciousness, but represented as belonging to past experience of thing, and so to be experienced. The thing is thereby truly *interpreted* for its feeling value. Not merely pain, as being experienced, is connected with thing, but as having been experienced, and to be experienced. Thus only arises that sense of the experienceable, that real *apprehension* for the future, which is so valuable an acquisition in the struggle for existence. Feeling quality comes thus to be assigned as real and permanent property of things, and every cognition comes to imply representation of feeling value, and so to be a basis for emotion. But all sense of experienceability is founded on sense of experience; the sense of things as possibilities of sensation and feeling is based on actual relatings of feelings to objects in simple direct experiences.

Fear is in itself pre-eminently a painful state, and we have to inquire as to the origin and nature of this pain. The statement of the problem in general form is, how does that which does not yet please or pain, but is only 101cognized as about to do so, give immediate pleasure or pain?

We have already expressed the opinion that fear is based on more than mere pain revivals; there must be true representation, the revival must be appreciated as representation of past experience, and indicative of future. The painful agitation consequent on prospect of pain seems, indeed, to include as pain element more than revival pain, but it is only seeming. Where does the pain come from which a person feels at the mere prospect of pain unless from the past? The pain is, of course, not the identical pain feared. Again, one cannot see how a cognition in itself, entirely empty of feeling, can cause a pain, except as acting as a link in a chain of association whereby conjoined past pains are revived. So far as fear is pain, it is, we may be told, revival, for representation of pain is not pain, and cannot cause pain. The pain which arises from cognition of pain to be experienced appears in a strict analysis to be wholly re-occurrence stimulated thereby, and not any new and peculiar mode of pain at pain. That this is the case is apparent from the fact that we can only have the pain of fear so far as we have experienced pain. Poignant pains experienced are the basis of poignant pain in fear. The knowledge that you are soon to re-experience an intense pain leads to an intense dread, in which the intense pain is revived from former experience. There are, to be sure, in the phenomena of fear in highly developed consciousness, complex pains which cannot be ascribed to revivals, reflexes upon consciousness of the great tension and agitation thereof, pain of loss of self-possession and self-power, and other modes which proceed from consciousness of consciousness, but this does not bear upon the question how mere cognition of pain, as to be experienced, can in itself give pain; how there arises from mere apprehension a pain which is more than and distinct from the revival pains.

102But, however we may be puzzled to see how mere cognition of experienceable pain develops a peculiar pain which is the essence of fear, yet we must acknowledge its production to be a fact. We may say, indeed, that the bare thought of pain even when conveyed by the printed word—the abstract sign of an arbitrary vocal name—is not without a tinge of a peculiar fear-pain which does not wholly consist of revivals. When preparing to go out into the storm on a very cold day I have pain in anticipation of the pain I am to receive from the bitterly cold wind. Now I may have preliminary shiverings, and there may be recurrent painful sensations as I look intently at the raging elements, pains which return from actual experiences which I have before undergone and at the time knowingly connected with wind and snow. But all these revivals, while the basis of my fear, do not give the distinct pain quality of the fear. The pain which I do experience when I actually step into the biting blast I know at once to be entirely distinct in quality from that which I before felt at the anticipation, the real pain, of fear. Again, when I say, "I was deeply pained to hear of it," and when I say, "The noise pained me greatly," I indicate that difference between purely mental distress and sensuous pain, between pain at representation and pain referred to presentation, which is to be emphasized in all our study of emotion. With a man in the hands of hostile Indians the tortures of fear are quite distinct in quality from the tortures actually endured. The agony of fear is a *genus* apart from the agony of physical pain.

Again, if the pain in fear were derived from revivals, then the nature of the pain in different states of fear would be as different as the sensations feared. But as a matter of fact the pain in fear of cold, fear of heat, of famine, of punishment, etc., is substantially of the same quality. I may fear one more than another, but the real mental agitation and pain which constitute the fear are in 103all cases essentially the same. If the pain in fear were sensation revivals, then fear of cold and fear of heat would be quite diverse and contrary in quality of pain value, but we all know that the dread of a cold day and of a hot day are in themselves essentially the same in nature. As far as the states are pure fear and have a pain quality, the conscious activity in both is entirely similar.

Further, if the pain in fear were wholly of revival nature, not only should we expect fear of different sensations to be correspondingly distinct, but we should also expect the pain in fear to never exceed in amount and intensity the pain feared as indicated by measure of past experience. But we know that our fears are often much more painful than pain feared and than our experience of past pain. The pang of fear, of sudden fright, is often more acute and intense than any direct pain we have ever experienced. The terrible convulsions of fear which we see in the insane give evidence of pain which could not have been reflection from direct experience. That excessive and sudden fear which turns men's hair gray in a few hours and transforms their whole physical system is plainly not any revival from the individual's past experience. As revealed by its effects it is often, perhaps, greater than the whole amount of pain they have ever suffered. Where, in the direct-experience form, pain is greater in the fear than the real pain suffered, we express the fact by the common phrase, "more scared than hurt." In all such cases the pain in fear is not the revival of past experiences of the object feared.

Fear is, in the main, the peculiar pain coming from consciousness of experienceable pain, but in general in all complex consciousness it is marked by dissolution and weakening of mental force. There is a shrinking of will, and a clouding of cognition, a general unsettling of all mental elements, a commotion or agitation which destroys the organic *consensus* of consciousness. But any excessive functioning of some element in consciousness, of emotion 104life, as fear, or of any other form, is unbalancing and detracts from normal activity of the whole. Fear, however, in its normal measure and form arose and was developed as a desirable stimulant; where it becomes paralyzing in its force, it is pathological in quality. Also where fear is pathologically intense it tends to disappear in sensation feared. Cognition becomes so weakened that sense of representativeness is lost, the thing feared is no longer brought before the mind in its potential quality, but is immediately apprehended as present in its influence—though really objectively absent—hallucination is produced, and fear naturally reverts to its earliest and direct form in immediate experience. As cognition is still further weakened the sense of object as giving pain is lost and fear in any form entirely disappears. The pain is not felt which before was feared to be felt. Fear thus in the general order of its disappearance repeats the order of its appearance and growth.

Fear always includes some sense of object. The apprehension of something evil to happen is the basis of all fear, but the thing, or, subjectively speaking, the objectifying, may be extremely vague. We may fear that some harm is to befall us, but what and how, we know not. We must suppose that in early stages this bare objectifying of approaching pain was a regular incipient form, that an indefinite fear preceded every case of defined fear. We, as a rule, attain a full objectifying with such ease and rapidity that this form does not often appear.

A complete fear movement, then with reference to cognition includes four stages: first, a very general sense of object as about to give pain; second, an increasing definition of object up to the maximum of clearness, thus marking the highest efficiency of the fear function; third, a decreasing definition of object till, fourth, a purely indefinite objectifying is again reached. Every fear, if it attains a normal life, will rise, culminate, and decline in 105this way. Even in man, where the full development of single simple psychoses rarely proceed undisturbed, there is yet observed a general tendency toward these stages. I awaken in the night at a sudden noise with slight and vague fear; suspicious sounds increase my fear and I listen and look more intently till I see clearly and quite fully crouching near the bed a dark body which I make out to be an armed burglar; as he approaches with his pointed weapon fear will most likely become so intense that I see less and less clearly, and a shot might terrify me into vague but very intense fear. If the object is discerned to be not a burglar but a chair, the fear quickly lapses. At a certain point of maximum clearness either a weakening or an intensifying of fear weakens cognition. Too much or too little pain is equally injurious to the knowing activity. Low psychisms examine and clearly define only that from which they have something to fear or hope.

The qualitative relation of the pain of fear to the pain feared varies greatly with the evolution of mind. Fear-pain could not have originated as a substitutionary function for the real pain except by being at the first somewhat less in quality than the pain to be endured, otherwise there would be no economy in the function. The progress of this function is to secure at less and less expense of fear-pain the suitable reaction. The function of fear being to escape a greater direct pain by a less indirect one, the progress of the function is in diminishing the amount of fear-pain for required effectiveness. The small original gain in the ratio is increased by small increments till in the highest minds proportion of fear-pain to pain feared might be represented by $\frac{1}{\infty}$. The pain in the usual fear which commonly induces me to step from the track before an approaching train, or which enables me after reading some advice on the subject to take precautions against the cholera, is evidently in infinitesimal relation to the pain feared. When fear is unsuccessful, as in anticipating a 106visit to the dentist, we, of course, suffer a double pain, both the fear-pain and the pain feared.

Often we must observe that the pain of fear is equal to or greater than the experience feared, and we have to ask how this disadvantageous excess could have been evolved. Often the pain of anticipation turns out to be far greater than the pain anticipated. However, a little reflection assures us that the excess of fear in many cases is only in appearance. We do not fear too much upon the judgment we have formed as to the coming pain, but we have by error of judgment assigned too much value to the pain. When a person being initiated into a secret society trembles with fear at being told to jump from a precipice, when he really is to jump but a few feet downward, his fear was perfectly just according to his judgment. If his belief is perfectly assured, the mortal fear will make him offer the most strenuous resistance and most likely secure his release from the ordeal. In all such cases the feeling is right enough, but the estimate of future experience is inaccurate. When an animal is terrified at its own shadow the fear is justly proportioned to the estimate of danger, which, however, happens to be erroneous. In the evolution of mind in the struggle for existence, more and more accurate calculations of possible injury are attained, and fear becomes more and more rational. Educated men fear only what is worthy of fear; they fear many things that lower minds do not, and do not fear many things they do. The true excess of fear is where we fear against judgment, as when, knowing the safety of travel by rail, I am yet constantly in fear while aboard a railway train. When

I still continue to fear, though I know the fear to be groundless, this is a true hypertrophy of fear. We constantly observe those who are fearful and timid against their own reason. When dangers known are compared with dangers obscure or unknown—and perceived to be unknowable—the fear of the unknown often prevails 107against the fear of the known, and we prefer with Hamlet to fear the ills we have than fly to others we know not of.

I must in conclusion express my conviction that while the physiological and objective study of fear and other emotions is of very considerable value, yet it is only introspective analysis which can reveal the true nature and genesis of fear and all emotion. What fear is and what is the process of its development can only be determined by the direct study of consciousness as a life factor in the struggle for existence. This I attempt in the present chapter, with the main result that fear, as indeed every emotion, does not consist of pain or cognition-revivals in any form, but is a feeling reaction from the representation of the feeling potency of the object.

CHAPTER VIII

THE DIFFERENTIATION OF FEAR

Fear, according to the analysis we have made, includes representation of object in its feeling value, predominant tone of mental pain, and will recoil. Fear in its primitive form, as we have seen, was a sudden and transitory phenomenon in consciousness, a simple thrill of feeling awaking will to spasmodic violent effort in the struggle for existence. All states of fear in early psychical history were practically alike in quantity, quality and intensity. Every fear is like every other fear in its pain tone and will effort. Every object and event considered as painful is equally feared; there is no distinction of more or less fear, nor any qualitative differentiation. Very young children manifest equal fear disturbance and seemingly identical in nature on all fearful occasions. Prospect of vaccination, of a scratch, of the pulling of a tooth, of a whipping, of an amputation, produce equally paroxysms of fear, waves of painful emotion, which discharge themselves in muscular contortions. The lowest animals likewise appear in all cases frightened to the same degree and in the same way. It must be said, however, that this period of simple undifferentiated fear is undoubtedly very brief, and embraces in the individual and the race but a comparatively small number of phenomena; but a careful study, even by the method of approximation will, I believe, show it to be a definite initial phase.

109While this primitive undifferentiated fear, which acts with the same force and quality in all instances, confers upon the organism which possesses it a great superiority over those which do not possess it, in the race for life, and thus marks a great advance in psychical progress, yet it is manifestly uneconomical in its action in that there should be precisely the same amount and quality of reaction in all cases. So when a considerable number of organisms had attained the power to fear, competition would inevitably lead to some differentiation, and this doubtless first in the direction of greater economy. The animal which could fear much or little, according to the degree of actual injury threatened, would have a great advantage in the struggle for existence over his fellows. The amount of pain in prospect is definitely gauged, and the fear pain becomes proportioned thereto, and so the will effort and muscular exertions. Fear in its earliest form sets the whole motor apparatus going at the highest rate, the whole organism is at the highest pitch of activity, and life and death struggle happens at every apprehension of pain, no matter how small the reality. Later, through discrimination, animals become capable of either a slight scare or a great fear, according to circumstances. The fear force is gradually rationalized and made less spasmodic and so more adaptive. The fear pain becomes proportioned to the real amount of pain and so to injury actually imminent.

This mode of evolution by decrease rather than increase of intensity may seem peculiar. Fear, however, certainly originates as a simple outburst of considerable strength relative to the individual organism, and the first step in fear growth is a development in the representation-of-object element in fear which tends to reduce the essence of fear as pain-emotion. Spasmodic primitive fear in becoming intelligent loses intensity in the essential feeling aspect. Other things being equal, the intensity of fear is 110inversely as the definition of its object. The dimly and uncertainly known is always thereby more fearful than the well known and familiar. However, as regards primitive psychism, we must remark that all phenomena are very large in relative quantity to individual capacity, but very small in absolute psychological quantity. A fear which convulses a very small mind would make but a very small disturbance in a mind of very great capacity. An amount of fear which would absorb completely one consciousness capacity, would require comparatively little force in a mind of greater calibre. The lowest minds are possessed by their fears, higher minds possess them, do not "lose their heads," *i.e.*, both cognition and will co-exist as stable controlling elements. Primitive consciousness is constantly at saturation point, phenomena occur only in linear consecutive order, and every phenomenon is a feeling-willing which absorbs the low conscious capacity. It may then, perhaps, be regarded that the evolution of fear is not through absolute decrease in intensity, but an increase of conscious capacity, whereby greater definition of object becomes possible and coincident with fear-pain of original quantity. The complete determination of this question must then await a fuller analysis, but the relation to individual capacity in the evolution of fear remains apparent. Whatever may be the absolute quantity and intensity of the fear phenomenon, its relative quantity and intensity changes very greatly.

The number of adaptive degrees of fear which are ultimately evolved and of which any very high mind is susceptible, is quite beyond our present means of psychological analysis. We have no phobometer to register all the gradations, other than the popular usage of language, but between "I was scared just the least bit," and "I was scared stiff," or "scared to death," there is certainly a vast number of intermediaries. Terror is an intensive term denoting strong fear, and a terrible fright is a redundancy 111for extreme fear. By the use of adjectives and various qualifying phases we roughly denote a number of fear degrees, but scientific precision is wholly lacking. Such expressions as "I have very little fear of him," "I fear him a little," "I fear him greatly," "I fear him very much," convey a meaning indeed, but no exact measurement is indicated.

Terror is often used as a term not merely for fear in general, but for fear which paralyzes by its force. The individual is often "rooted to the spot" by terror, he loses all power of motion and becomes as an inert mass. With animals even of the lower grades this is doubtless often a pathological manifestation. We find that predatory animals are often furnished with apparatus to inspire benumbing fear in their victims. Various means, as inflation of size, strident noises, etc., are employed with great effect. On the other hand, we find that predacious animals seek to reduce the stimulus of fear in their victims by quieting and alluring methods. Both hypertrophy and atrophy of fear are disadvantageous, and we should see then in paralyzing terror an instance of over-development of useful function which produces the direct opposite of the normal fear. Fear, the great means of salvation to all weaker organisms, is also in its highest intensities taken advantage of by enemies. Hence the due graduation and restraint of fear becomes one of the most important lines of mental evolution for the organism preyed upon, but the over stimulation or undue weakening of the fear function in its prey becomes a most important object and advantage for the predacious animal. This evolution is often by the individual disadvantageous variation when this is advantage to some other organism; and, as living beings are soon divided into the two classes, those who flee and those who pursue, the destroying and preserving of the chief psychological defence becomes a leading form of psychic growth of a pathologic character. Fear in its 112origin was certainly a stimulant to action and not sedative. However, so far as fear effects an unconscious mimicry of death it often reaches thereby negatively to conservative action, and paralyzing fear is thus explained by the general law of advantage in the struggle for existence. We can then trace a double evolution of fear, on the one hand as leading to action, on the other to inaction, but the former will, I think, be found to be the primitive form. The primary and main function of fear in all life is in a duly modulated energizing in view of approaching injury, and the depressing mode is secondary and exceptional.

Again, we must remark upon the sense of personal weakness, or, objectively stated, the sense of overwhelming power, as entering into fear. I cannot agree with Mr. Mercier that this is a mark of all fear. In its origin and early gradations fear, as we have noticed it in the immediately preceding paragraphs, requires no other cognition than that of pain to come. Self-measurement of power in relation to that of pain giving object is certainly too complex to be primitive, nor do the simplest forms of fear as we observe them in ourselves and judge of them in lower organisms pre-suppose any such process. Primitively every perception of painful event fills consciousness with the impetuous self-conserving fear revulsion. There is neither time nor capacity for estimating one's own strength or weakness in relation to opposing power. By the very low intelligence only the immediately imminent is apprehended, and action is always immediate, short, and decisive. In fact, it is now probable that originally painful events are really actualized by the mind, and the fear is thus at the event as actual, rather than as ideal, as represented as to be. Certain it is that mind, in its hurry to get ahead of natural harmful agencies in their action, must in its earliest pre-apprehensions have no room or time for dynamic interpretations.

Of course the whole value of sense of one's own superior 113power is in fear, thereby securing the contingency of the painful event, but sense of contingency upon one's own efforts no doubt first occurs at a considerably advanced stage, much beyond that of simple fear. Primitively mind regards events as being, or about to be, with no sense either of their certainty or uncertainty. Early mind cannot appreciate certainty, for it knows not uncertainty, it has not yet accomplished the prevision to which certainty and uncertainty may attach; it cannot say, "I fear this will happen," or "I fear that will not happen," but only "I fear or do not fear the thing happening, the event coming." The world of the earliest psychical life is simply factual, and the fears are simple and wholly undifferentiated. Fear certainly antedates the perception of contingency and of one's own agency in producing contingency. Even in the ordinary fears in human consciousness sense of personal power in relation to pain-giver is actually subsequent to the fear phenomenon and reacts upon it, but is not constitutive of it in its first impulse.

Fear is first graduated by the increasing discrimination as to the amount of pain and injury to be inflicted, and later it is graduated by the sense of the painful event as more or less contingent, either in the natural course of things, or as determined by the individual's strength in warding off impending evil. Taking chances and risks is learned, and becomes often very advantageous. Fear is also greatly diminished and modified by acquiring a sense of one's individual power in overcoming or resisting pain given. The rabbit, often chased by a clumsy dog, evidently fears him less and less. Man, both by his increasing knowledge of natural contingencies and by his increasing power over elemental and animal pain-giving forces, fears less and less. The inevitable evil, sure to come, and sure to overcome, is that which strikes intensest fear, as we often see in criminals led to execution.

The discrimination between the animate and the inanimate 114also differentiates fear. When this distinction is fully achieved, the attitude of mind toward each in fear is plainly distinct. The thing, perceived as having psychic powers, and capable of purposive evil and self-directive of its movements, awakens thereby a complex of feelings which rapidly develops beyond our present powers of analysis to follow them. For the present sketch of the early natural history of fear it is sufficient merely to remark this differentiation as one of prime value in the struggle for existence.

However, as we have before suggested (p. 106), the nature of fear, purely in itself considered, does not depend on the nature of the object feared; thus fear of cold and fear of heat are perfectly alike as psychic facts, though having regard to very diverse physical facts. Animistic mind, indeed, reacts to all objects differently from naturalistic mind, yet in its essential quality fear is identical in both. In fear of a storm, both as a purely physical manifestation and as the expression of the psychical nature of a deity, the fear act is by itself quite the same; the fear pain and the willing are quite the same, but on the more external, the representation side, they do greatly differ, the complication being greater in the latter instance, and introducing a complex of feelings. Fear in the narrowest sense does not reach to the object to consider its nature, to regard its objective quality, for this is the base of very different feelings; but fear proper is engrossed in object purely for its immediate pain significance; it is given up to viewing personal pain infliction. I am inclined to think, then, that we shall find that mind is primarily neither animistic nor naturalistic. The only interpretation of object which is first made is as pain or pleasure given, and a personalizing and impersonalizing stage is decidedly later. We must remember that mind at first goes only so far as it is positively obliged to by the struggle for existence; and hence, though it is quite impossible 115for us to fully realize such a simple state, yet originally objects were discriminated merely as pleasure and pain sources. Object at first was of the more vague sort, merely an indefinite *locus* for pleasure-pain;

something painful or pleasurable is the discrimination, but attribution of sentiency or insentiency is not yet reached, for no interpretation of the sort is yet imperatively demanded. It is so ingrained in us to perceive beings as either living or non-living, that it is quite impossible to thoroughly conceive a state so primitive as to be unable to rise to this attribution or distinction. However, like the bare statement of a fourth dimension in space, the statement that pre-animistic mind exists or has existed, a way of looking at objects entirely without reference to their personal or impersonal quality—this is intelligible, and hypothetically required by a complete theory of the evolution of mind. In a *dolce far niente* of perfect sensuousness, even the adult man sometimes approximates this stage, and the actions of very young infants are best interpreted as expressions of a similar state. Things for them seem entirely uninterpreted and unperceived, except as imparters of crass sensual pains and pleasures, as mere pleasure-pain potencies.

A very important differentiation of fear is brought about by the extension of the time sense. Fear begins with a *minimum* of time sense; only the immediately impending, the absolutely imminent danger, suffices to awaken fear. But in the struggle for existence the advantage of being influenced for action by the more and more remote, in time, determines a rapid extension in time to feared events. With man actions are thus influenced by fears, which reach even beyond the present life. The cautious and prudent are those whose fears are far-sighted, and who, conducting themselves accordingly, maintain supremacy over the short-sighted and improvident. *Carpe diem* is, from the point of view of 116evolutionary psychology, the cry of the retrogressive fool.

The time differentiation of fear is recognised in popular language in the term—dread. I am frightened in the night by a sudden noise; I am alarmed for the safety of a child awaking near a precipice; but I dread next week's task. Of course dread, like other popular psychological terms, is plastic, and often denotes fear in general, and is often used intensively, or to denote vague fear, still it is the most correct and distinctive term for fear of a more or less remote event. It would be most interesting to investigate the relation of distance in time of feared event to intensity of the fear, but we have as yet no standards for estimating in mathematical ratios either time or intensity psychologically considered. It is not, of course, physical determination of time as minutes, hours, etc., with which we are concerned, but only with variations in sense of nearness or remoteness of event. Our sense of time is most variable, and fluctuates from many causes, so that hours sometimes seem minutes, and minutes at other times seem hours. However, there is, doubtless, other things being equal, some fixed relation between our sense of the nearness and remoteness of a fearful event and the intensity of the fear, but we may well doubt whether it can ever be reduced to any law of inverse squares like that of physical intensities. A criminal sentenced to die at the expiration of thirty days certainly has a marked increase in fear as time approaches, or rather, as he has sense of the time approaching, but a quantitative analysis is beyond our present powers.

A most important but tolerably late differentiation is the altruistic form of fear—fear, not of others, but for others. Psychic life is at first wholly self-centred, there is no perception of things or interest in them otherwise than as bearing on the experience of the self. Other selves are wholly unrecognised, and pain-giving effects 117to them are then unperceivable. In very young infants we see a close approximation to primitive selfish life. The exact point in the history of life when altruism is developed by the struggle of existence is not at present determinable, but we may well believe that it arose with the evolution of the sexes in separate individuals. Fear for mate and offspring is obviously an essential advantage in the progress and perpetuation of the kind. Pure altruism is not at first attained, and there is only the faintest gleam of appreciation of pain-states in others, and genuine feeling therefor. The sexual appetite is, like other appetites, purely selfish at first, and the animal fears the loss of what will satisfy in an individualistic way, quite as he fears that food may be taken away or destroyed. Even in higher psychisms much that we readily interpret as altruistic is often mainly personal; it is not a true regard and emotion at pain and injury imminent to others, a manifestation of feeling at their experience as such, but mostly a feeling for their experience only so far as it involves our pleasure-pain. When sociality and interdependence of organisms is attained as a great advantage in the struggle of life, when personal experience is perceived as dependent upon experiences of others, then a feeling value attaches to the experienceable for others, yet selfishly at first. Even parental oversight and care must originally have been selfish—the satisfaction of a personal craving, rather than the promotion of the well-being of another, considered for its own sake. Real and pure altruism must, indeed, be accounted, even in human society, as a rare phenomenon, perfect self-forgetfulness being almost impossible even for the most developed consciousness, owing to the strength and persistence of an indefinite heredity of selfishness. Fear for others is, then, in truth, merely an indirect fear for ourselves; and particularly so is this true in all lower consciousness. But we must acknowledge that elements of real altruism do enter and 118do grow in value and strength in the evolution of consciousness, and we must, if we adhere strictly to the principle of personal advantage as determining evolution, find a reason here for a singular and seemingly incompatible manifestation. Regard for the good of others is not always indirectly regard for personal good, and self-sacrifice is certainly an element in psychic life, even in lower consciousness, where we often seem to see a distinct struggle between egoistic fear and altruistic fear, as in animals protecting their young. But we see the same in an animal defending food from being acquired by its enemies.

Advantage for the race is certainly gained, but this wholly unconsciously; and it plays no part in the actual psychism of the individual. In a highly social, which is also in the most effective and advantageous mode of life, it is certain that the purely self-seeking will be at a disadvantage in general, whereas those who give themselves up to help others are by others so helped, that the final *status* of the individual is higher and better than if he had been wholly a self-seeker. However, he who, perceiving this law, sets out to be altruistic for his own ends, invariably suffers defeat in the long run, for entire disinterestedness can alone avail. But the problem of altruism, from an evolutionary point of view, cannot here be further remarked on; a fuller discussion would lead us too far afield. However, we are convinced that altruism springs up and grows like the other elements of psychic life, as functional in the largest way to the demands of life in the struggle for existence.

Horror is a distinctive term for altruistic fear. When on a train, I am *terrified* if I perceive a collision imminent and inevitable, but as a mere spectator walking near the tracks, I am *horrified* by the prospect of a collision. One may be "in mortal terror," but not in mortal horror.

Our sense of the feelings of others towards us, whether they be egoistic or altruistic, determines a large class of 119reflex emotions which are often very subtle. If we perceive that some one is fearing us or fearing for us there is immediate reaction on our part. Feeling response to feeling acts and reacts in a multitude of complex ways, as we cannot but observe when in the company of very "sensitive" people. The "sensitive" one is he whose emotional life is governed by his perception of the feelings of others toward himself, and he becomes wonderfully responsive to the least expressions of emotion toward himself. The delicate responsiveness of women, their intuitions, are merely quick perceptiveness of emotion expression. The fears of such are largely concerned with this dependence on the emotional attitudes of others toward themselves; they fear to incur displeasure, they fear loss of love, etc. Thus psychical phenomena become more and more determined by psychical phenomena as interpreted and considered with reference to the self. Panic is contagious fear, and has originated and been developed as securing mutual safety in societies of animals. However, there is less real fear on occasions of panic than is often supposed, for much of the expression which we read as fear inspired is really merely imitative, and does not signify any real basis of emotion. Moreover, we must note that there is no direct contagion, but the perception of fear in others merely leads us to dimly body forth some fearful events as impending, which representation involves the full phenomenon of fear. There is also a discrimination as to those who shall impart fear; the fear of a child on shipboard will not start a panic, while the fear of a captain would. Convinced that there is something worth fearing, we fear, and make frantic efforts to escape.

We have before mentioned (p. 89) the peculiar fear of fear. The latest and culminating differentiation of fear is awe, and the highest, most refined development of awe is in the feeling for the sublime. The sense of magnitude and mighty potency of injurious agents or agencies in 120themselves considered, and not as immediately affecting the individual or any individual, is the essential element in awe as a species of fear. This fear is then neither egoistic nor altruistic, but impersonal. We fear neither for ourselves nor others in standing awestruck at the foot of Niagara, but a sense of overwhelming greatness and might stirs a thrill of emotion which is at bottom a sublimation of fear. The view which to a peasant or an animal would give terror, or produce no emotional effect whatever, with very rational and sensitive minds produces awe. Awe does not, as early emotions and fear generally, lead directly to will, it is not a stimulant to action, and thus has not been evolved by the principle of usefulness for action which governs the general course of physiological and psychical evolution. It is evident that with awe and the sense of the sublime emotion has a value and end in itself. In the higher evolution of man we see that the psychic elements evolve no longer in a strict dependency for their value in securing advantage and success in the struggle for existence, but comfortable existence being practically assured, psychic development is pushed on in lines ethical, emotional and intellectual, for no practical end, but for their own intrinsic value. Thus the feeling for the sublime is a purely independent development, which, indeed, is based upon man's capacity to fear egoistically and altruistically, but is really exercised solely for its own sake. A consciousness which has had no common fear stage, could never arrive at awe. We stand in awe of persons who are totally beyond us in their superiority, who exist in a sphere of power and glory, which transcends even our understanding, and thus awe has a religious as well as æsthetic side.

The chief differentiations then of fear we note as intensive dread, as altruistic horror, as impersonal awe. The chronological order of evolution may be denoted in this order—fright, alarm, terror, dread, horror.

CHAPTER IX

ON DESPAIR

Despair is a phase of painful emotion which is certainly related to fear, yet is very distant from it. Despair has always a fear basis; we can only despair where fear is implied, and what does not excite fear will give no hold for despair. I must first fear a pain before I can despair of escaping it. The prisoner condemned to death must fear death before he will be in despair at the prospect of it. Yet while despair always implies fear, fear may often exist and that in very strong form without despair. The prisoner often displays great fear, but no despair.

There is, in fact, a strong contrast between fear and despair. Fear normally stimulates effort, despair depresses it. Fear is active, despair passive. Deep dejection and lassitude mark despair, while fear is intense agitation and activity. Fear in its original and normal function is stimulant of defensive action, fear as paralytic being secondary or abnormal, but in normal despair there is absolute inertness. Fear, again, in contrast with despair, is direct and transitive. I fear the pain or injury, but my despair is only in relation to it, despair *of, in* despair, etc. Fear is at the evil itself, it is a direct attitude of mind toward it, through an ideal pre-experiencing, the very representation of any pain as experienceable carrying with it a thrill of fear. But despair concerns itself, not with the pain *per se* as experienceable, but with the inevitability of 122the painful. Fear rests upon idea of pain, despair, upon idea of its inevitability. "I despair of escape," means a recoil of painful emotion at inevitability of painful experience. Sense of complete and permanent inability to attain an end, whether release from pain, or positively, a securing a pleasure, generates commonly this distressful emotion. Despair is not then simple pain at pain, but at the unavertibility of the pain. Despair is then the mind bent down and crushed by the sense of the inevitable and irremediable nature of the pain, positive or negative, it experiences or is to experience. Despair is, indeed, hopelessness, though all hopelessness is not despair. There is no hope in stolidity or in stoicism, psychic modes quite distinct from despair, but which take the place with some natures.

Again, we must note that while fear has its degrees, and may be but partial, despair is always complete collapse. I may fear a little but not despair a little, I may be frightened "just the least bit," but not despair a little bit. The hostess who is "in despair" because the ice cream has not come, speaks truly, however, for the affair is for her so important and momentous as to be the basis of real despair. That which is the occasion of despair must always be or seem of capital value.

An adjacent and often precedent state to despair is desperation, which is a feeling of the almost inevitable. In the face of heavy odds there is often awakened a painful emotion which we term desperation, and which leads to strong and furious will action, to an intense and general struggle which is often advantageous. An enemy fears to drive his adversary to desperation. In desperation we take one chance in a thousand or in a million; for example, the leader of a forlorn hope. It would be difficult to say whether despair or desperation contains more of pain, but they are obviously quite opposite in their character. To combative temperaments and with pugnacious 123animals the sense of the seeming inevitable is often stimulative of desperation rather than despair. Such are "game" to the last. A criminal of this type will run amuck rather than submit to his fate in despair. The desperado is defiant to the end. With some whose natures are balanced between reflection and action there are in the face of the inevitable or almost inevitable rapid fluctuations of despair and desperation.

Dismay is another form closely akin to despair. Dismay is the immediate result for feeling of a sudden cognition of great difficulties and pains as imminent. As the transition stage of rapid movement in feeling toward despair, as the sudden falling in temperature from hope, it is really incipient despair. Dismay is essentially temporary, and settles quickly into despair or rises into renewed hope. Though but such a passing mode, it yet has for the moment that sense of self-efficiency annihilated which is so characteristic of despair. Consternation is very intense dismay.

But what now is the real quality and inner nature of despair? what essentially is this strange drooping before inevitable loss, injury and pain? and what is its significance for life? Despair is certainly a very advanced and complex emotion, and we can do no more at present than merely remark on some of its most striking features.

A most noticeable and remarkable quality of despair is its introactive tendency. When the whole strength and vital motive, of a full-grown teleologic psychic life—the *dilettante* is not capable of despair—is suddenly and completely withdrawn, there results, not indifference nor *ennui* but a deep disturbance which is active on the *minus* side of mental life. The complete breaking up of great and absorbing hopes and of the free objective activity flowing from them brings will tension down, not simply to *nil*, but gives it a spring back into the negative region beyond the line of mere quiescence and 124indifferentism. Despair is a revulsive process by which the whole mind is broken up, just as a propeller wheel running at high speed out of water or an engine working at high pressure when disconnected from its shafting, tend to wrench and shatter themselves. Desire is not really extinct, but latent; though smothered it burns inward. This is that peculiar cankering, corroding quality, which is always so marked in despair. Will, not self-shattered, but forcibly pent by external circumstances, gives a sullen restlessness to the mental life now turned in upon itself. Hence the capacity for despair will be directly as the co-ordinate capacity for action and reflection in any individual, and as such co-ordination marks the highest level of conscious life, despair is certainly a phenomenon of exceptionally complex and advanced consciousness.

Again, we note that despair is intensely and oppressively a pain state, but the dull despair pain is distinct from racking fear pain. What now is the nature of despair pain, and what the reason for its peculiar quality? Here is not as in fear a feeling pain at pain, but at the idea of its inevitability and completely destructive power. The actual pain foreseen may seem bearable and excite little feeling, but it is the total loss of personal success, the complete thwarting of self-realization, that moves the mind to despair, that causes that sickening, dull, emotional pain which we term despair. Thus despair is eminently a disease of self-hood, an egoistic distemper, the strong and large individuality being peculiarly subject to it. However, the general problem of despair pain is practically the same as of the origin and nature of fear pain, which has already been discussed. Whether any mere representation induces pain, and how it does so, is certainly one of the most difficult problems of emotional psychology. We have in a previous chapter sought to indicate in a general way that purely subjective or mental pain which is not in any wise revival of sensation or objective does really 125exist. Also since pain *per se* is always simple and identical, the differentiation of pains as seemingly quite different in kind, as fear pain, despair pain, etc., is really due to sensation, will, and other elements which closely adhere to pain and give it a certain local colouring. The whole emotion is a complex of various factors which are closely knit into a single state which to common observation seems simple, but which is really constituted in its *ensemble* by the total specific forces of many elements. In psychics, as in physics, we know that common sense analysis of phenomena must be at fault, and that one who says "I certainly have an entirely different pain when I fear and when I despair," is as much in the wrong as he who maintains essential diversities in material substance, or radical distinctions of species in the organic world. So we must believe that the peculiar quality of the pain in despair exists, not in the pain itself, but is really the colouring result from various coincident sensations and ideas. The lowering of the mental tone far below the zero point is greatly accentuated by refluent waves of organic sensation set up from the physical basis of the psychic disturbance.

How, we may now ask, did despair ever evolve and become a well-defined psychic form? in what way in the course of natural selection could such an apparently disadvantageous variation have arisen and been developed? The serviceability of fear is plain to every one, but of what possible value could despair be in the struggle of life? The one who gives up in despair is but very rarely doing the best thing. If we cannot look to the general principle of evolution, serviceability, how can we account for the appearance and growth of such a phase as despair, except as abnormal variation, a disease, profitable to the enemies of the individual, and so developed by and for external organisms. As there is an abnormal pathological variation of fear, which we have previously noticed, and which is forced in its development by enemies who profit by it, 126so despair is a psychic disease, entirely hurtful to the individual, and, so far, only advantageous for its enemies. Despair is, without doubt, one of those altruistic variations which serve, not the individual, but some antagonist in the struggle of existence. To bring one to despair is to make him entirely helpless and wholly at our mercy for our own ends. The possibility that active-reflective natures may prey upon themselves is thus stimulated into an actual phenomenon whose growth is continually fostered by those whose advantage it is to reduce the individual to a helpless condition. Despair is hardly an hypertrophy or atrophy of any normal tendency, it is rather a pathological *genus* by itself. The capacity for despair being inherent in the general formation of mind as subject to collapse, it arose solely in response to the needs of organisms warring upon the organism afflicted. The whole field of physical and psychical altruistic variation under the general law of natural selection, decadent and self-injurious characteristics being stimulated and maintained in a kind of artificial selection, is an interesting but unexplored field, attention so far having been turned to the individually advantageous as determining element in evolution.

Despair is a disease of advanced and mature psychic life. Children are, in general, incapable of despair. It implies a well-developed sense of self and a general experience of the world. High and strong emotional natures, but rather weak-willed and narrow of intelligence, are predisposed to it. Occasions which would lead to despair will with lower natures be unnoticed or lead merely to stolidity; while with the highest natures, there comes heroic endeavour and wide searching for means and methods.

CHAPTER X

ANGER

In studying any state of consciousness we first inquire what constitutes its dominant factor; if this is sense of object, we call it a cognition; if effortful action, it is a volition; if the marked feature is pleasure-pain, we term it a feeling. Finding that the consciousness is a feeling, we would next inquire whether the pleasure-pain is mainly determined in its colouring by direct presentation, and so is a sensation, or whether this dominant colouring comes indirectly through representation, and is thus what we term an emotion. For example, the distinction between "I feel a pain in my shoulder," and "I feel pained at your conduct" illustrates the most radical division of feeling. If emotion is founded on an appreciation of the experienceable, which has developed under natural selection, we must look upon the emotional power in general and upon the various emotions in particular as merely advantageous psychoses which are as clearly determined by general evolutionary laws as the merely physical organs like heart, lungs, wings, horns, etc. It is clearly desirable that the organism should look before, should anticipate experience and so direct its way; but bare anticipation has no value in itself unless it powerfully stimulates will through emotion. All conscious life above the most primitive is eminently and increasingly anticipatory, and so becomes more and more infused with emotional powers. Among the earliest developed of these 128in the struggle for existence are fear and anger. The fear group, embracing large numbers of allied forms, simple and complex, has been discussed in a general way in the preceding pages, and we now come to some consideration of the correlative anger group.

The *rationale* of the evolution of anger is not far to seek. We have seen that fear is the spring of defensive action, and it is obvious that anger is the stimulant to offensive action. Fear is regressive, anger aggressive. Fear is contractile, anger expansive. Fear is the emotion of the pursued, of the prey; anger the emotion of the pursuer, of the predacious. Emotion in the service of life evidently has two great psychic ramifications from this point, and the whole world of emotion-beings, which compose the greater mass of organisms, is hence divided in two great divisions, a fear class and an anger class. Likewise in relation to opposing natural forces as to neighbouring competing and destroying organisms, the same distinction is to be made according as the animal either combats or flees. Shyness or fierceness, timidity or irascibility, these are characters which divide the animate world into two grand antagonistic groups. Zoology has recognised this psychic differentiation as a marked and essential feature in its nomenclature, thus *lepus timidus*. In fact, the most important part of evolution is the psychical; in this, indeed, lies the whole significance and value of the organism. The attainment of more and more advantageous psychic qualities is the main trend of evolution, for psychic power as distinct from main force, like that of the elements, is far and away of the most value in the struggle for existence, and ultimately, as in man, it achieves the subduing all lower powers, natural, vegetable and brute, to its own ends. It is psychical quality, moreover, which determines physical, and not *vice versâ*. Thus it is not the possession of claws, fangs, etc., that makes an animal fierce, but it is fierceness which 129develops and maintains these weapons of offence. Thus it is, though thus far practically overlooked by scientists, that psychic development, especially on the emotional side, is of the utmost importance as the prime factor and motive in organic processes. The central core of life is emotional capacity, and this in its evolution determines the whole external morphological trend of evolution of organisms which is so closely followed by the science of to-day. But the science of the future is comparative psychology, which, when once placed on a secure basis of interpretation, will determine the real and inner law of evolution as a psychic movement incarnating itself in a succession of animate forms. But a sure method of knowing a psychic fact as such when it occurs, and what, how, and why it is, is yet to be discovered and applied, and extra-human and even extra-ego consciousness is a field, so far, for little else than hypothesis. If this remark be turned against us, we say that our work is mainly a deductive interpretation of the course of psychic evolution from the general standpoint of natural selection, reinforced and illustrated by introspective investigation, and merely using the most obvious facts of comparative psychology in a very general and provisional way. We do not profess to show where, how, and when mind originated, or what particular powers any certain organisms possess, but we do endeavour to show how the principle of utility may be made a key to the study of a very perplexing region of mental life—the emotions. We proffer then merely a very general sketch of the history of emotion as a life factor, hoping that it may, at least in its general scope, be of service to future explorers. In taking up this subject of anger we do then thus reiterate the position we occupy and the method we follow.

Anger like fear certainly originated at some critical point in some individuals life as an advantageous variation of essential value. A vital issue at some early point 130in the history of life determined the genesis of this new psychic mode and function as a stimulant of aggressive will action. Very likely it was in competition of organisms for food that some favoured individual first attained the power of getting mad and violently attacking its fellows, and so obtaining sustenance. However this may be, certain it is that a direct attack is often more self-conservative than attempts at escape when injury threatens; it is a greater advantage to destroy pain-giver than to shun it. Fear enables organisms to avoid loss, but it does not accomplish positive gain, as anger does through overcoming hindrance. Anger is often also more economical for the forces of the organism, and thus, in general, predacious animals are longer-lived than even those of their prey who may attain a full length of life. Even in the face of great odds a direct attack is often more serviceable than attempt at escape. Anger is certainly the primitive motive force of all offensive action, though of course we cannot say that the animal got mad because it saw the serviceability. Psychic evolution, at least as far as new powers are concerned, never comes by teleologic foresight, and, indeed, cannot by the nature of the case. The animal did not definitely set out to get angry because it foresaw the value, yet in the earliest angers there must have been effort, a certain *nisus* which marked the new form as a real attainment, a marked achievement. That the provoking occasion gives rise now to anger inevitably and naturally, that anger comes upon us and overcomes us is true enough, but in its earliest phases anger must have been, like other just evolving factors, supported only by powerful will effort. The oftener the early psychism got mad, the easier it got mad. Facility came only by practice, and a large variety of occasions, besides the simple critical and original one, were gradually utilized by the anger faculty. But in its original form and occasion anger was, no doubt, akin to 131that we see when an extremely timid animal at the last extremity will turn in anger and fiercely fight for its life. Such an attempt, sometimes successful, marks an origin of a new mode of conscious emotion which may never return to the individual again during all its future life for lack of occasion. If often returning and often improved, a definite new habit of emotion is established, and from being a fearful animal it may at length become dominantly irascible, and so belong to a totally distinct psychic genus.

By the evolution of anger then, as in contradistinction to fear, two grand divisions of animate existence were set apart, two great psychical orders as fundamentally distinct and important for evolutionary psychics, as invertebrate and vertebrate for biology. The rise of the back-boned animal is not more important for physiological morphology than the evolution of anger for psychical morphology, and, indeed, as we have before remarked, the psychical growth is ever the broadest and deepest fact in evolution. By the acquirement and predominance of the anger stimulus certain animals became differentiated as a distinct class from their fearful neighbours, and they then by this new impulse gradually attained instruments of offence, and also by increase of size became physically distinct forms. Henceforth the animate world becomes divided in a more and more marked way into pursuers and pursued. By mutual interaction fear is increased on one side as anger increases on the other, and the division into timid and fierce, predacious and prey, becomes more and more established and marked.

We take it then that it was a most momentous day in the progress of mind when anger was first achieved, and some individual actually got mad. If the exact date and the particular individual were ascertainable a memorial day set apart for all time would not be too great an honour. In the struggle of existence, other things being equal, the 132most irascible is the most successful, faring the best, securing the best mate, and having the best and most numerous progeny. Susceptibility to anger becomes a necessity to a large class of organisms, and those who will not get angry and fight for their interests are surely trampled on or pushed aside to become starveling or outcast.

Is now this primitive anger an absolutely new power, a *de novo* evolution, or is it possible to study its rise as a gradual differentiation from some other factor? Must we not view psychical evolution like all evolution as coming under the law of continuity? How then explain the sudden rise of apparently new and distinct forms like anger or fear? Anger as a response to the demands of life seems from the very first to be as distinctly and peculiarly anger as at any time in its development. The peculiar quality which makes anger anger, does not seem to appear as a gradual differentiation from other elements as slowly emerging from previous modes, but we can only judge that it bursts suddenly upon the field as a new and unique creation, which does not find its explanation in pre-existent forms and cannot be traced as a gradual evolution from them. On the other hand, while it does not at first sight seem possible to regard anger as being from the first other than a radically new power and activity determined, indeed, by the struggle for existence, but wholly unexplained in its essence and formation as a consciousness related to and differentiated from other consciousnesses, yet we must acknowledge our profound ignorance of the real morphology of mind and what is the real nature of mental differentiation. Here the problem is altogether more difficult than in biology, where the appearance of new forms like wings can be readily traced as slow modifications of previous members, the physical possibility and *rationale* of which is easily seen to be inherent in the physical constitution of the body and its circumambient matter, the air. However, in the present state of our 133psychical knowledge it is quite impossible to attain any similarly clear conception as to the formation of new psychical forms. We may see why they should be called into being by the necessities of animate life, we can perceive their functional importance from the first, but to trace their morphological development as gradually assuming their peculiar qualities as modifications of already existing activities, and as inherently possible in the psychical constitution of things, this is clearly beyond us at present. We can conceive that the earliest anger was weak and rather ineffective as compared with the fully developed anger of later life, but we cannot see that it was any the less anger, any the less purely and wholly *sui generis* than the very latest and strongest form. Has it ever in its earlier stages that hybrid and mixed character which marks it as a modification of existent factors? It is certainly not a modified fear, to which it is, indeed, a polar opposite.

But we may perhaps regard anger, and fear as well, as modified from previous general emotion. We may, indeed, consider it likely that some general emotional phase preceded the special emotions, just as a general indefinite pain and pleasure preceded definite pains and pleasures. It may be considered as probable that emotion first appeared as a purely undifferentiated disturbance sequent on sense of the experienceable pain, this general emotion being neither fear nor anger, but the basis from which both develop. The psychic agitation we term emotional very likely began in a purely general form, yet it is hard to understand how peculiar forms develop therefrom. We are too far from such inchoate experience to readily come to any appreciation of its method or mode. We may be disturbed as to something imminent and know not whether to fear or be angry, but this in general means only a rapid alternation of fear and anger according as the mind runs back and forth between fear and anger-provoking 134elements. It is unlikely that we can trace in any such a purely undifferentiated emotion.

At the best we but throw the difficulty farther back, for emotion *per se* is then the *de novo* form to which the principle of continuity does not seem to apply. If anger is a traceable modification of some more general emotion as combined with definite representation and volition modes, yet how the peculiar anger quality is achieved is still unexplained. On the whole it seems simplest and truest to assume the first impulse of anger as a perfectly new and diverse wave of emotion suddenly generated in answer to some extreme urgency in the struggle of existence.

The analogy of organic and psychic evolution may be pressed to a certain extent. It is plainly possible to set in order an evolutionary series of light—sensing organs, eyes—from most elementary to most complex, and it is quite as possible, though yet to be done, to set forth in similar genetic order a series of psychic states as offence-sense, *i.e.*, angers, in their increasing differentiation. But previous to any eye, to local visualization, there is a period of common sensation when an absolutely simple organism is in every part equally responsive to light; in a crude way the whole organism reacts to light, from which stage by traceable specialization the eye as a light-sensing organ is gradually developed. Here analogy would seem to fail, unless we consider it to be the stage when any psychosis, *e.g.*, anger, requires the whole consciousness capacity, mind being merely a capacity for the recurrent but isolated single-activities. Mind certainly but slowly grows into that sum of organic coincident interdependent yet distinct consciousnesses which we commonly think of under the term, mind. Anger in its very earliest and lowest form is no doubt an absorbing naïve isolated wave, as common to mind as a whole, that is, as making up the whole of mind for the time being, is perhaps 135in its measure an analogy to common sensation. Anger may then be but a common emotion, answering in a certain aspect to light-sense, sound-sense, etc., as purely common sensations. But we must remark that general sensation is not to be confounded with common sensation, or general emotion with common emotion. Common sensations are, indeed, usually very general in form, and a sensation *per se*, a purely general sensation, is probably very rarely anything else, yet when we close the eyes and direct them toward the sun, the general sensation of light we receive—very like the original primitive common sensation—is general, yet by a special organ. The word common refers, not to the special nature of the function itself, but the fact that the function, whether special or general, is performed indifferently, or practically so, by the common whole. A sensation of coloured light is more special than a mere sensation of light, and this than mere general sensation of force, but all may be accomplished either by common sensation or special sensation. General emotion may similarly be either common or in organic co-activity. There was certainly a time when consciousness existed which was not and could not be anger or fear or even an emotion *per se*. Pre-emotional and pre-representative consciousness was so absolutely primitive, general, and common, that psychology as a necessarily automorphic science will be very long in coming to any understanding of this field, but yet we must set it off as something which must always receive some consideration. Anger is not a property of all consciousness by the nature of consciousness itself, but is merely a possible mode dependent on circumstances for its development at a certain psychic stage.

What now is the inner nature and what the constituent elements of the anger state? Comparatively few reflect upon their emotions save from an ethical standpoint, and very few indeed attempt any analysis of them. To determine 136the process and exact psychical constituents of getting mad and being mad, may seem to many a quite useless and foolish introspective endeavour. If a person is angry, he is angry, and that is all there is of it, will be the general verdict of common sense. You can dissect flowers into their parts, you can analyse rocks and soils, but any emotion such as anger is wholly unanalyzable. No one can know what it is to be mad until he has once been mad, and, thereafter, he can only reflect upon it as a peculiar excitement, a powerful agitation, whose occasions and results may be fully traced, but which in itself is *sui generis* and irresolvable. The form of consciousness we know as being angry, is really a simple wave of emotion which stands by itself as an elementary and ultimate form.

Suppose we acknowledge these remarks as true, we may yet maintain that anger, like all emotions, is a highly complex state of manifold factors whose sum total, whose grand resultant, is a seemingly simple and peculiar *status*. Why should one arrangement of atoms produce a peculiar perfume, another a peculiar stench? Anger may likewise be merely an unexplainable *ensemble* of early ascertainable elements.

Certain it is, in the first place, that sense of object is necessary to anger. One cannot be mad without being mad *at* something. The attitude of mind is objective, and even rage in its blindest moment preserves this attitude. Blind with rage, means no more than that various definite qualities of the object are lost in the intense emotional reaction at pain-giver. At its height, anger preserves, indeed, only the barest apprehension of object; but this is intense and overpowering in connection with the sense of it as infringing and injuring. In the transports of rage and fury, the movements are wild and reckless enough, but always antagonistic, implying outward destructive activity. Anger is the fixation of the mind upon some object in its quality of personal hurtfulness, and 137is revulsion, not *from* it, as fear, but *against* it. With early psychisms, all perceptions of objects end in either anger or fear, and a large part of early education consists in learning what objects to be fearful of, and what to be angry at. The alertness of wild animals is determined mainly by either nascent fear or anger. When a dog is suddenly wakened from sleep he generally shows either fear or anger. This is merely an illustration of how the dimmest sense of object immediately connects itself with emotion as primitive and fundamental tendency. The organism perceives the object, and representing its imminent hurtfulness, feels fear and dashes away from it, or feels anger and dashes against it. These are the two simplest possible reactions with sense of the experienceable injurious. In fear there is elimination of oneself from the injury, and in anger the elimination of the injury from oneself. With later anger and fear these processes of elimination themselves become matters of representation, and make a large part in highly-developed forms.

A knowledge which very generally enters into anger is the comparative estimate of power. A cat scratches us, we are angry; a lion threatens us, we are afraid. The progress of the lower psychic life is largely in learning what is best to fear and what should excite anger. That which at first angers will often, when better understood, produce fear, and *vice versâ*. Wild animals at first often show merely anger when molested by man, but soon manifest fear as they learn to appreciate his superior power. The African elephant learns to distinguish between the savage with his spear, and the white hunter with his rifle, and is merely irritated or angry with the one, while he manifests genuine fear of the other. The young of animals and of man continually show irrelevant fear and anger. They are generally either over fearful or over irritable. Our own feelings are powerfully modified by varying estimates of opposing force and injury. If, in passing through a dark 138street, I am tripped by what I take to be a child's snare, I am angered, but upon noticing that it is a fuse to a dynamite bomb, I am thrown into intense fear. In general, any sensation, as of sound or light, in its lower grades of intensity produces anger, in higher occasions fear. As a rule when reactions induced by either fear or anger are uniformly unsuccessful, natural selection favours the development of the other.

While the comparative estimate of opposing force with one's own is general ingredient in anger, anger being fear-limited, it is not, as Mercier would indicate (*Mind*, ix., p. 346), a constant element in anger. We often see cases of anger, and have perhaps, ourselves, experienced anger which is totally unrelated to a sense of power. Some animals seem at times utterly fearless and utterly unaware of the tremendous crushing force they angrily oppose. It is, moreover, altogether probable that anger and fear originated and received a certain measure of development before any capacity of measuring comparative force of antagonist arose in mind. However, the discrimination between overwhelming and slight force is certainly tolerably early, and is obviously a very necessary factor in self-conservative action. Yet it is very unlikely that this was an element in primitive fear or anger, which must have been no more than a simple emotional reaction to perceived injury without any reference to whether pain-giver is more or less strong than pain-receiver. The earliest fears and angers of infants seem to be quite devoid of any guidance from sense of powerlessness or power, but merely direct, unthinking reactions.

A marked and constant element in anger is hostility. This is the aggressive fighting attitude of will which is exercised toward and against the perceived pain-giving object. Anger can never subsist without this volition element, and it always appears as direct simple reaction to anger-provoking object. Anger always exhibits itself as 139hostility, openly and freely in lower life, and in higher life, which is often disingenuous, the hostility as real psychic act remains, though somewhat concealed in physical manifestation as long as angry mood exists. The will tendency is always toward the violent removing and destroying of the offending object. However, naïve primitive anger does not include in its hostility giving pain for pain received, making the object suffer in turn, which is, indeed, far removed from the capacity of primitive mind to conceive. Anger in its earliest form does, of course, inflict pain where its object is pain-susceptible; but this, it may confidently be said, cannot lie in the intent of the pain-inflicter. The simple original ebullitions of anger do not include intent in any form. Volition is powerfully and directly incited by the emotion without the intervention of any idea. The only representation in the simplest anger is the representation of pain experience impending which occasions the excitement, which then directly and violently starts will-activity; but the representations of destructiveness and pain-infliction as ends become guiding ideas only in the slow evolution of anger toward more intelligent forms.

Pain is certainly a prominent element in anger. This pain is the emotional pain, the pain at pain, whose nature and origin we have commented on in the chapter on fear. The mere representation of pain to be starts a violent pain quite distinct from the fear-pain, yet like it, pre-eminently central and subjective. Precedent, however, to both fear and anger-pain, is the simple pain which immediately arises on representation of pain, the prospect of pain being immediately and peculiarly painful in itself. This commonly continues throughout, and gives a dominant pain tone. But there immediately succeeds a rush of either fear or anger emotion, each intensely painful in opposite ways. The pain which results from the anger, which is by the anger occasioned in me, is again distinct from the pain 140in and of the anger. Anger is itself a state of pain. In its earliest forms, as rarely and with difficulty attained, there is still another pain connected with anger, the pain of exertion and stress. But all the pain factors, as more or less continuous, make anger, as emotion in general, a complex pain state. Thus, when angered by a man shaking his fist in my face, we trace first a purely subjective pain at prospect of pain, then a rush of aggressive emotion which embodies in it a pain of its own, then a pain which reacts from the peculiar tension of the anger state. Of course, in our stage of evolution, anger has become such an inwrought factor that it arises spontaneously, it overtakes and overcomes us, not we reaching it; and so the stress or labour pain is absent. It is never or very rarely an effort for us to get angry, but it must have been for our very remote psychical ancestors.

While it may be said with truth that some people are never so happy as when mad, yet we must remember this does not alter the fact that anger is radically a pain state. There may be a pleasure from anger excitement, and from successful anger; there may be a pleasure in the mere exercise of aggressive power; but the happiness meant is mostly the excitement pleasure *plus* the delight which always comes from freely following out one's nature. Especially when the outflow of natural force in an irascible man has been pent up and restrained for some time, a fit of anger is altogether a delightful experience, the pleasure of relief in a habitual function. Thus an occasional fight is necessary to the pugnacious amongst both animals and men; it is an inbred function and tendency which must work itself out, or render the being as miserable as a rodent kept from gnawing. But all this does not interfere with the analysis of anger as fundamentally painful. Happiness is a very late evolution, and, as the reaction from freely working out one's strongest tendency, it is unfelt by early minds, which only gradually attain 141inwrought tendencies and so the capacity for being happy or unhappy. To witness a fight is likewise to a large class of minds a supreme felicity. This is largely the pleasure which comes at second hand from representation of participancy. And so, to have a fight described, or to read about it even, is a source of considerable representative pleasure to many, a spurious and reflected anger, and an ideal fighting in the fray. However, all this leads far away from primitive emotion, which is now our main concern.

We may grant then that sense of the object giving pain, sense of comparative power, hostility, and pains of various kinds, are usual elements in anger; yet it is evident that anger is explained by no one or all of them. It is not a mere aggregation and mixture of states, it is essentially a compound which has in some unexplained way a peculiar quality which is not in any of its constituent elements. When I am angry, there occurs a phenomenon which, while based on and inclusive of these factors, is yet peculiar in itself. The flush of anger, the wave of emotion, the tempest of passion, bases itself on and includes cognition, hostility, and pain; but it is more—it is a deep psychic disturbance of a peculiar and undefinable kind which we recognise when we have it, but which we cannot analyse. We express the nature of anger metaphorically, indeed, when we speak of an angry man being "hot," "boiling with rage," etc., as opposed to being chilled and frozen stiff by fear. The being angry is obviously a kind of being pained at pain quite opposite to that of fear. It is also true that I may see threatening injury, I may be pained, I may combat, but not be angry. There are other and higher motives which may bring about the violent will offensive activity so often required in the struggle of life; but we may take it that anger is the most primitive, and throughout the whole range of psychism the most common offensive motive, and so of the utmost importance as a life factor.

142 Which shall we regard as the more primitive, anger or fear? Were animals at first universally timid, and subsequently acquired anger as an advantageous variation, or was anger the first, and fear the complementary and later evolution, or may we suppose that they developed in strict correlation? The earliest manifestations of emotion with some animals, and with some human infants, seem to be anger. Everything perceived to be painful irritates and makes them mad, and they are quite fearless in the presence of overwhelming danger. These but slowly learn to fear; by hard experience they learn the hurtfulness and inutility of combatting in many cases, and occasions which would once make them mad now cause them to fear. On the other hand, we observe many of the very young who seem to be universally fearful, and but slowly acquire "spunk" and spirit. Mental embryology thus, at least with our present very imperfect knowledge, is quite indecisive on the question. If fear and anger were wholly determined by relation of predacious and prey, then we might suppose correlated simultaneous origin; but we know that obstacles and injuries, not from competitors, but from elements, forces, and objects of nature, were the first environment and the first field for struggle. Organism began as a weak thing planted amongst manifold opposing forces, where fear was quite the most salutary emotion and anger useless. If, as we must deem probable, mental function in general and emotion in particular reaches back toward primitive organism, it is likely, on merely general grounds, that fear is the more ancient and original emotion, though anger was closely subsequent. The general conditions of life at the first would demand the development of fear more imperatively than anger. Certainly, however, both emotions are sufficiently primitive, as is shown by their being so ingrained and dominant forces in the whole range of lower psychic life.

All higher animals, moreover, are peculiarly sensitive to 143 and observant of signs of anger and fear. Rarey, a most excellent judge, made it an axiom of his method that horses are extremely acute in detecting either fear or anger in those who deal with them, and this is also noticeably true of animals in general. These are also the emotional attitudes which are earliest interpreted by children. Now what is soonest, easiest and surest interpreted by psychisms above the lowest may be taken to be fundamentally primitive and such are fear and anger. To discover with readiness and certainty the emotional states of organisms about them, because these states are the motives of very important activities, is clearly an advantage early gained in the struggle of existence. It means preparedness, and there is a nascent anger to break forth against the fearful, or fear or counter-anger prepared against the fear discerned or suspected. The inter-related activity of these two emotions is the chiefest and most interesting spectacle we see in all lower psychic phases.

But we must notice now a form which seems on the whole to belong to the anger group, and that is hate. Hate often precedes and succeeds anger, and the object of anger is peculiarly apt to be the object of hate. The man whom we hate very easily angers us, and he who provokes us is one whom we are apt to hate.

Yet a person may be very provoking, even exasperating, and not be hateful, and *vice versâ* for hate. It is obvious then that while the object of anger and hate is apt to be the same, yet it is viewed from very different standpoints, and the emotion reactions are somehow very different. "I hate him," and "I am angry at him,"—these expressions denote very distinct emotions. While anger and hate are both aggressive emotion reactions against the pain-giver, yet in their nature they are essentially diverse. In general we hate him who deliberately and constantly provokes us, who establishes himself as a deliberate enemy. It is harmful, 144opposed intent that particularly stimulates hate. But anger is most generally a sudden flash of feeling leading to violent repulsive effort against pain-giver, but without any insight into intent. The immediacy of reaction is accomplished through anger; but hate, having more of insight and foresight, is more slowly generated, and is not so directly and promptly active. I may be angry at one who casually pinches me in sport, but I will hate him who continually pinches me in spite. I may be angry at the child who in its childish play often interrupts my studies, but I do not hate it; this I reserve for the malicious boys who continually put tick-tacks on my windows. And so also inanimate things often arouse anger; but we hate only the animate, and then mainly when we discern deliberate, purposed offence. To be sure we often hear some such expression as, "I hate the very sight of that house"; but here the term hate denotes loathing, and is only a little less flagrant misuse than when I say "I hate ham, but love beefsteak."

Hate, then, marks in a very noticeable way the growth of psychic responsiveness. A prevision of psychic attitude of others, especially the emotional and volitional, is of the utmost service as helping to and preparing for an appropriate response. Thus we may believe that quite early in mental evolution there came an appreciation and interpretation of the psychic modes of others as affecting the interests of the individual. We may judge that this is probable by the very apparent difference of reaction of even certain of the lower animals in the presence of threatening dangers from common material things, and from animate beings capable of being not merely crushed or pushed away, but intimidated and frightened away. Young children learn quickly to distinguish between mere physical events and psychic expressions, and to feel and to act toward the psychic in the peculiar manner which will best serve them. Thus it becomes of very 145definite value to excite fear in enemies, but even a low animal learns speedily that it cannot terrify a large stone which prevents access to food. Now fear and anger obviously do not specially belong to the rather advanced class of emotions which are always psychically responsive, for, in earliest phases at least, both fear and anger may be taken to have no reference to the psychic quality of the object, but only to the physical quality as painful and injurious. However, later fear and anger become cognizant of the psychic attitude and responsive thereto; but it may be said that hate from the first is a psychic responsive, it is an answer to the psychic attitude of others as interpreted by the individual as turned towards itself. Hate is always against evil intent; anger and fear may be. Hate and anger are both intensified by hate and anger in the object—though this may often occasion fear—but fear, on the contrary, is greatly weakened, and sometimes turned into hate or anger, by perceiving its object as fearing it. I naturally hate those and am angered with those whom I perceive as having the same passions against me; but he whom I see fearing me does not thereby inspire my fear for him, but tends in quite the contrary direction. Yet mutual fear in equally matched opponents is consistent with mutual anger and hate. Fear, with those who are capable of inflicting about equal losses on each other, acts as a check upon anger and hate, and gives caution and wariness to passion itself.

The object of hate then differs from that of anger and fear, as being invariably a psychic quality in another as injurious to one's own interests. Injuriousness *per se* does not excite hate as it may anger and fear. Animals, indeed, often seem to hate that which has no psychic attitude toward them, and may be wholly incapable of it; but this is error of judgment, just as we ourselves often find ourselves wrong in hating where we supposed there was evil feeling toward us, but where we now see there is 146none. Hate disappears the moment we discover our mistake of interpretation.

While hate often views its object very largely from the retrospective side, as opposed to fear and anger, which are generally prospective, yet hate originally must have applied to the present or latent potency of the object for harm, for only in this wise does it reach self-conservative value. In early psychic life there is no time or place for purely retrospective emotion like revenge and resentment. Hate is not essentially a paying back for the past offence, but a will-inciting emotion of immediate, or imminently prospective value. In fact, though we say, "he has done me injury and I hate him for it," yet we do not hate the dead injurer or the one so crippled as to be entirely powerless against us. Certainly there is no value for our interests in injuring the one who is past injuring us, and from the self-conservative point of view to exercise ourselves in hate or anger in such a case is to waste energy. Feeling for what has been done against us, purely as such, is plainly sheer waste of force. The past is irretrievable, and emotion about it is valuable for life only so far as the past implies the future. Thus it is that hate, arising because of self-conservative value, and developing under natural selection, never becomes wholly retrospective.

Hate then is at first much the same in its elements as anger. It is always objective. Hate is always of something, though extreme passion dulls perception, yet at its normal tension hate, like other emotions, is incentive to beneficial cognition. We are closely observant of those we hate. Beside sense of object, there is the will-stirring, the hostility, which is prominent in anger, though here more controlled and not so impetuous and naïve. Hate thus often allies itself with fear, but anger is very rarely coincident with it, though there may be rapid alternations. There is also a hate pain which is a parallel complex to the anger pain already analysed. We might term hate a 147distilled anger, and yet this signifies little, for the innermost emotion seems very distinct. Like fear and anger, hate seems a *genus* by itself, and in its essential feature as emotion-reaction, quite beyond scientific analysis, which can point out its conditions, but not account for their total value or for the peculiar quality of hate disturbance by which hate is hate. Hate can be appreciated only by realization, but no matter how long we reflect upon and try to catch its exact nature in some definite formula, the essence of hate always eludes, and presents itself as only a bare simple psychosis wholly indefinable and inexplicable in its essential nature.

But if we turn now to the origin and development of hate, shall we arrive at anything more satisfactory? Is hate a modified anger, or is it from the first a wholly distinct emotion and not slowly differentiated from any preceding psychosis? Hate evidently belongs with anger as aggressive emotional reaction, but it is very hard to see how it could originate by any slow growth, and it seems easier and simpler to regard it as being a unique response to some very pressing demand in the struggle of existence.

The whole subject of mental differentiation needs clarifying. Are we to consider mind merely as a sum of many distinct modes each of which has, in the course of evolution, appeared suddenly in answer to the demands of life at a critical period, and is faint, indeed, yet from the first having a distinct and peculiar quality by which it suitably stimulates will, and that the sole growth of these diverse forms has been in intensity and by various associations with other states? or are we to consider that mind was originally a very general vague state, which, by a continuous and traceable differentiation, has slowly developed into many different modes? Certainly the latter seems the more rational. To conceive that there are no essential and radical subdivisions in mind, that not even knowing, 148feeling, and willing, are fundamentally primitive, but each, and each form of each, but modifications of precedent modes, this is a theory which is enticing in its simplicity and in its analogy to physical evolution from a single underlying material element. But when we come to particular investigations, as this of the origin and development of hate, we cannot well discover any modes intermediate between it and say, anger, which are the links in a continuous evolution, but for aught we can see or conceive, hate is as much hate the first time it appears as at any subsequent time. The links in the evolution of mind from phase to phase are all missing, and how are we to supply them? Of necessity as subjective facts they must first be realized, before they can be known, but how can this be done by a consciousness which has long outgrown them? We cannot discover these fossil and extinct forms objectively, as the paleontologist discovers extinct species, but in some way we must re-enact and re-experience them in our own consciousness before we can know anything about them. If every mind embryologically passes through the several stages of its general evolution in the race, still the strange intermediate forms which might then have existed are beyond the recall of the reflective stage, when we first demand to know the history of mind. And when we appeal to comparative psychology we are equally in the dark, for we must judge animals by ourselves, we can interpret their consciousness only by our own, and they may have very rude and peculiar forms which are unknown and unknowable by us. Thus the limitations and difficulties of subjective research are especially brought up to us in evolutionary study which thus seems wholly confined to *a priori* speculation. While we can conceive it likely that hate was suddenly brought into full being by the demands of life, yet it is hardly a rational view of emotion to regard it as a *per saltum* series of 149distinct psychical species called successively into being by the exigencies of existence, which indeed, is a view almost as ultra-scientific as that which regards all mental modes as direct endowments from Deity.

But though on general scientific analogy we are led to believe in fossil mental forms, in missing psychic links now extinct as regards our own consciousness, but which were the germs of our present distinct emotions, perceptions, etc., how are we to discover and investigate them? Can we work our own consciousness back through the multitudinous stages of its past evolution, through myriads of human and pre-human forms to the confused, primal, undifferentiated psychoses?

Certainly the forms which lead up to such an emotion as hate and from which it is gradually evolved must be realized, must be actually felt in some measure before they can be understood and analyzed. Here then seems a great barrier to introspective evolutionary psychology, perhaps insuperable, for how can mind retrace itself, involute itself, in the interests of science? Mind is fundamentally action, motive-feeling, which, in connection with cognitive forms gradually achieved, becomes from mere pure pleasure-pain a very complex manifold. We feel many of these forms in our own experience, and we can say of some that they are the higher, of others that they are the lower and more primitive. Thus fear, anger, and hate are generally regarded as low action-motives as compared with love of truth or justice. But while we distinguish in our own consciousness and by analogy in the consciousness of others a considerable variety of psychic forms, they are, so far as we are able to see—and we have given some special attention to this in discussing fear and other emotions—invariably distinct, and each has its own peculiar quality, and we do not find, and we should not expect to find, the intermediate forms any more than the 150anatomist would expect to find in man a radial starfish structure. The hazy, indefinite phases which mark evolving consciousness into new forms have been long done away with for such emotions as hate, and it would seem an impossible task to ever bring them back. When we let consciousness lapse of its own regressive tendency—and undirected consciousness tends always to revert to *wild* states—we fall down through a series, but it is by steps, and no gradual descent, that is, defined mental forms succeed each other, with no transitional phases which are both as differentiating into either. We have mixed states, indeed, but these have no evolutionary value in this line, being merely coincident distinct psychoses, and not an intermediate differentiating mode. The psychoses which we call lower and which we naturally *fall into*, were really a higher level once for some remote ancestors, and it was only by occasional great efforts that fear, anger, hate, etc., were reached, by just such efforts as now are required by many a worldling who would be religious and would attain a feeling for holiness, or that of a Philistine, ambitious of reaching æsthetic feeling, who endeavours to appreciate the refined, elaborate power in a poem by Rossetti, or the simple human grandeur in a painting by Millet. In some forms we know what it is to try to feel, to have dim and vague stirring of æsthetic emotion, and to reach new levels in emotion generally, and we know the stages of differentiation and the severe *nisus* of the earlier realizations. On the *nisus* side of our psychic life there is abundant opportunity for every one to observe the process of mental differentiation, and how slowly evolving a new emotion is, for instance, before it reaches a definite form, but there is the great range of purely natural, spontaneous life, deriving its whole *impetus* from ancestral minds, where, as in hate and anger, it is impossible to study the slowly modifying forms precursory 151to the distinct mode. How can we find or produce in ourselves a state which is not yet hate, but merely hate in becoming, a half-differentiated, half-evolved hate? If we could put ourselves on the *nisus* side, and look up to hate as something to be reached, instead of something we may fall into, we might attain some idea of its process of formation. But since hate, anger, and so forth, invariably come upon us and overcome us, how can we appreciate their evolutionary stages? If we could trace these old intermediate disused forms which merely lead up to others, we should find them very strange, and should need an entirely new nomenclature for them. But to reach back and realize long outgrown and fossil psychoses, will, if ever possible, require more exertion and ability than even the intense struggle of the actual psychical advances which adds, by the efforts of exceptional individuals—"geniuses"—new modes of cognition and feeling to the mind of a race. To regress beyond a certain point is harder than to progress.

How then hate developed from non-hate, from anger, or from any other emotion, is obviously a very difficult problem. It would seem to us in our present stage of mentality that the first hate phenomenon was definitely and inexplicably such. We cannot perceive or conceive how the origin of hate is other than a sudden apparition of a new and elementary emotion in response to an extraordinary call upon some extraordinary organism in its life career. Yet we may easily believe that the direct occasion of its rise and progress was as complement to anger. Anger is certainly in general a very advantageous self-conservative factor, but by reason of its violence it requires a vast amount of vital energy to accomplish its end, and it thus also tends to disturb the cognitive power in its clear and cool actions. A burst of passion, though it may succeed in destroying the injurious, is both uneconomical 152and unintelligent. It is also a very transient phase. Anger will be defeated and supplanted in the evolution of life by some factor which has not these incidental disadvantages. Hate is such a superior psychosis, and is surer, steadier, and more economical than anger, and defeats it in the long run.

Hate then may be taken to exemplify the principle of antithetic evolution. We are careful not to raise the anger of some men and of some animals, and thus anger, or the capacity for anger, serves them as advantage and defence. We fear to make them mad. However, the antagonists of many individuals, knowing the weakening effect of such a strong emotion as anger, and knowing also how apt the angry one is to "lose his head," purposely stimulate anger to their own advantage, and the disadvantage of the angered. Thus, cunning and wary animals, impelled by hate, often tease and torment their stronger and larger adversaries and competitors into a furious rage, which is so rash and unintelligent that they are completely at the mercy of the weaker. Where in such a way as this an advantageous variation is turned into disadvantageous by an opposing form, as anger by hate, we have what may be called an antithetic evolution. New psychic variations are then continually stimulated by the earlier, and it is only for a short time that any variation maintains itself as purely beneficial, but an answering one soon takes advantage of its weak points and turns it from self-conservative into self-destructive. Under the constant success of opposing factors, there is gradual loss of value and soon disuse, with the inception of some new form to combat more effectively the opponent. This opposing form meanwhile attains dominancy, culminates, and is gradually ousted by some variation which has been attained in order to meet the new weapons on the other side. Thus, in the battle of life, offence and defence, attack *versus* retreat and 153counter-attack, mutually stimulate to a series of new and higher antithetic psychic variations.

The so-called problem of evil is, then, tolerably easy to a thorough-going evolutionist. All developments, all perversions which are self-destructive rather than self-conservative to the individual, have received their original stimulus from other antagonistic individuals to whose interest it is to promote these evils to the utmost. What is an evil to me is first so much of a good to him whose interest lies in defeating and destroying me, and he will take advantage of all my weaknesses to his own profit. Competition and struggle involve the existence of evils to individuals who are conquered and maltreated in the battle of life. Disease and death itself is necessary to evolution on a finite sphere. As long as the good and desirable is limited as compared with the number of those who want, competition must exist, and this competition must be by both cultivating advantageous variations in ourselves, and also by cultivating the disadvantageous variations latent in our enemies. Thus, evil sown in others that our own good may be advanced is the general law of all life. To injure as much as possible all those who oppose, and to get as many as possible well affected towards us, and to be subservient to our ends, this is the meaning of psychical evolution in all its earlier, and most of its later, course. On any scheme of evolution by struggle, evil to particular individuals is a necessary fact. We throw, then, the problem back to how and why life arose and developed through this competition mode; and all science at present can say is that it is the "nature of things," an expression which covers ignorance and is really metaphysical.

In all its later stages anger, and likewise hate as well, and all the allied emotions, attach only to what is distinctly known as animate. The futility and self-destructiveness of anger against the inanimate and insentient comes to be 154fully recognised. But early anger is quite undiscriminating. The hunter, who, pursued by an enraged bear, scatters his clothes and accoutrements behind him for the bear to tear in pieces, takes advantage of the unintelligent anger of the bear for his own ends. Since animals do not wear clothes they have no conception of what they are as independent insentient things distinct from the wearer. To the bear the weapons and clothes dropped by the hunter appear not as inanimate beings, but as living, vitally-connected parts of the creature pursued. The error arose, not from senselessness, but from lack of range of experience, and it is akin to the error of the ancient Mexicans who, having never seen a horse by itself, regarded a man on horseback as a single creature. A dog, the first time he sees his master unclothed, is greatly puzzled, and but slowly learns that clothes are something the master has and not what he is. When weapons, clothes, etc., are at length distinguished as property, there is yet a natural and right impulse to destroy them as injuring the owner; but the animal which stops to do this commits an error of judgment, as it is usually of more importance to despatch the hunter than to destroy his implements. It is the tendency of anger to destroy all which is in any wise connected with its object. This is true, not only of the animal world, but also of the lower human development. A savage in a fit of fury will slay, not only an offending fellow, but also his family and relations, and also destroy all his property. The uselessness, not to say the injustice, of such an indulgence of anger is only recognised at a comparatively late stage of evolution. Anger in its later form concerns itself only with purposive offence in its object, and vents itself solely on the individual offending. A clear distinction is drawn between animate and inanimate. Thus, my dog, playing with another, hurt itself by running into a tree, and gave an angry growl; but noticing the real nature of the paingiver 155as, not the other dog, but an inoffensive tree, his attitude immediately changed, and he seemed to take the injury as a matter of course. A puppy would in like case senselessly continue its demonstrations of anger to no good and perhaps to its own injury.

As to the function of anger and hate, this has already been intimated in the remarks we have just made on its origin and development. For function it is which gives rise to organ and activity; in some unknown, mysterious way the pressing life-struggle for useful mental activity determines ultimately its appearance. We know that extremely hard conditions, which would threaten the continued existence of animate life as a whole, or of any large subdivision, would give rise to new perceptions and emotions by which a saving remnant would escape; and on this principle we must expect the most signal psychic advance of the future at that seemingly remote period when mankind will be threatened with extinction by the slow refrigeration of the earth. A long-continued uniformity of easy conditions of life, as in the tropics, is distinctly unfavourable to psychic progress; but let a glacial period invade that zone, and the changed conditions would awaken such a struggle for existence in all organisms, man included, that new organic and mental types would be developed. The necessities of existence and the self-interest of the individual in an unceasingly sharp competition develop slowly in the few those mental modes which, from their functional importance, become the heritage of a race and *genus*; and these "sports" thereby secure to themselves a certain temporary dominancy. This is the history of life in general, and of man in particular. How demand determines supply, how necessity is the mother of invention, is obvious enough in man, who, clearly conceiving the function, sets about by his knowledge of means to accomplish the needed improvement; but in the lower life, which is incapable of such 156teleological foresight, we can only say that through pain of lack in the altered conditions of existence there is stimulated a blind, intense struggle, which, moving out in all lines, somewhere, at sometime, by mere chance hitting on the right variation, sticks to it and accomplishes its own salvation, and leaves descendants who tend in the same direction. New psychic qualities, as well as new physical organs, are in some way gradually determined through struggle which is practically blind. That mental variation, that bodily variation, which was incessantly demanded in the struggle of existence does somehow ultimately appear, is, indeed, a fact which, for the present at least, we can only state in this indefinite, unsatisfactory manner. Blind, pain-impelled will, fiercely striking out in every direction, does at length, achieve those new psychical and physical forms which are most needed by life. The chance serviceable variation is fixed and continued by reason of its serviceability; but when its utility wanes by reason of new life factors appearing or new conditions of existence, it is lost by disuse, or survives in rudimentary forms.

The function of hate is, like anger, to injure and eliminate the injurious; but what anger accomplishes by a sudden volcanic outburst, hate accomplishes in a slower, but surer and more subtle way. Hate is, as previously pointed out, a manifest improvement over anger as a method of offensive warfare. Other things being equal, the best hater is the most successful individual. Dr. Johnson had reason on his side when he said that he loved a good hater. A strong hater, who pertinaciously assails and injures his enemies, strengthens his own position and makes the largest place for himself in life. Hate, as a permanent, economically aggressive motion, marks certainly a great advance, and is of the highest import for life. If now hate has its own function as direct stimulus to offensive action toward those who will be injurious, 157toward those who are capable and likely to pain and harm us, how shall we explain the hate—and we might say anger as well—which arises at mere remembrance of injury, and which seems to have no immediate value for life?

In the first place we may well doubt whether any purely retrospective emotion exists, at least in early psychic life. The past, of course, has no value in and by itself; it is irretrievable, and emotional force spent upon it as such wasted—"no use crying for spilled milk." It may well be that for simple psychisms the past never exists as such; at least, it is never a stopping point, but a mere *datum* for interpreting the inexperienceable. The sense of experience, especially in its temporal aspect, is very difficult of analysis; yet we may say with some confidence that at first it does not imply a sense of either the past or future as such. The mind is immediately impressed by the injuriousness of the injurious, which, though coming, of course, in terms of the experienced, is not relegated thereby to a past time, nor is it at all dwelt upon as such for emotion reaction. Primitive emotion is not backward looking; for this is in itself entirely futile, and primitive life depends for its existence and progress upon utility. The value of emotion is in stimulating preparedness for defence and offence. The representation of injury inflicted comes up to early mind as some injury being inflicted, or imminently so, or is applied at once in interpretation of the experienceable, with no thought or emotion for it as merely past fact. Advanced psychic life may stop at the first step, may indulge in retrospection for its own sake, and not for its immediate value in understanding the experienceable, but primitive emotion is ever an alertness and anticipatory readiness.

If, now, we turn to some classification of the anger group in itself and in its general relation to emotion, we obtain something like the following:—

158

Emotion.	Reaction to injurious.	Regressive—fear.
		Aggressive—anger.
	Reaction to beneficial.	Receptive.
		Appropriative.
Anger	Simple anger or wrath.	
	Intensive—Rage or fury.	

Incipient—Displeasure.

Mild—Irritation.

Response to purposive injury—Hate.

Altruistic—Indignation.

Sentiment—Indignation and Hate.

Retrospective—Resentment.

Revenge.

Sub-hate—Detestation.

Despite.

Scorn.

But few remarks need to be added to elucidate the outline. Exasperation is plainly a late form of anger. It belongs to the period when anger has been subjected to will restraint, and when something passes all bounds of forbearance—is "perfectly maddening"—we are exasperated. Anger of a high and peculiar intensity produced by special and repeated provocation is known as exasperation. For intensive hate there seems no special word, at least, in English, though we denote it by adjective as bitter, malignant, virulent. Detest sometimes means strong hatred. Malice is not an emotion; it is a state of mind which is implied in hate, namely, deliberate intent to injure. We do not say we feel malicious; but if we hate, we are malicious. Malice is merely an objective term for a will element in hate, and denotes character of act.

The sight of injury done to others produces indignation. When law or principle injured and violated excites indignation or hate, we have that feeling for the abstract—rarely 159pure—which is termed sentiment. He who is indignant at injustice and he who hates sin have risen to the highest evolution of the anger group. For an account of resentment and revenge see chapter on Retrospective Emotion. In the earlier stages both anger and hate are rather undiscriminating as to rank or *status* of opposing object, but in later evolution there must be a sense of equality. When we consider the offending ones as entirely below us, as unworthy of our anger or hate, we detest or despise. Our relations with them may compel us to notice them and to have some feeling toward them, but we would not lower ourselves to fight them. To detest is to feel a strong revulsion, but it also in measure has a direct objective movement. Still, although detestation, despising, scorn, contempt, are by no means so actively aggressive as the other members of the group, they have evidently a direct affiliation with hate and anger. In all these there is direct repulse of all relation with what is below us, a position holding off and looking down upon the offending object as too small and mean for us to seriously oppose.

We cannot at present elaborate more fully an analysis, a genetic investigation, nor a classification, of what must appear to every attentive student of mind as a most important and extraordinary group of psychic phenomena. In all the lower psychic life with every perception comes an emotion reaction, very generally either of a fear or anger character. Everything perceived has a definite life meaning, nothing is indifferent, and, in fact, primitive perception cannot exist except as prompting and being prompted by emotion or feeling. For the low psychism there is no such vast collection of practically indifferent objects, a world of things, as maintains a constant and large place in advanced psychism. Lower mental life is piecemeal, inconsequent and broken, and wholly directed by feeling phases. Every object has its place only in relation to self-interest, as favouring or injuring. This is 160impressed upon those who have made any study of lower human types, and of wild animals, where your very presence, no matter how accidental and really meaningless, is construed as suggesting detriment, and suspicion is aroused, a preparatory stage to some fear or anger exhibition, one of those being often nascent, though sometimes not very active owing to the lack of full certainty as to your injuriousness. For the savage, who is incapable of disinterestedness, and wholly given up to self-seeking, the missionary and scientist must have some hidden personal motive, some intent to take advantage of them, and profit by them. From the first they are regarded with fear, anger, or hate. The strange and peculiar is hated merely for being unlike the self, and all non-conformity means personal slight and insult. With primitive psychism all objects are coloured by a strong emotion light, and this remains a tendency till the latest stages of evolution.

Anger and hate have by no means spent their force, even for human evolution in some of its more advanced forms. We all recognise the necessity of "spirit" to success. The one who is incapable of anger and of venting it powerfully is a weakling, and will be trodden under foot in the battle of life. The high sense of personal honour and advantage, which will brook no insult with impunity, or allow no injury to go unpunished and unresented, is still the *sine qua non* of worldly success. Show anger, hate, and defiance to all those who invade your rights; stand up and fight the battle of life against every oncomer, and secure and hold the position against all competitors. In the natural course of events—the struggle for self-conservation and self-aggrandizement—the meek do not inherit the earth, but rather those who are irascibly aggressive.

The most notable revolution in human history against the general course of evolution which we have been considering has come from Christianity. The world says, "If any one smite you on the cheek, hit him between the 161eyes"; the Nazarene says, "Offer him the other cheek also"; the world says, "If any one takes away your cloak, fall upon him and despoil him of his all"; the Nazarene says, "Give him your coat also"; the world says, "Hate your enemies"; the Nazarene says, "Love your enemies, bless them which curse you, and do good to them that despitefully use you." The law of natural evolution by fear, anger, hate, strife, is replaced by a new law of a spiritual evolution through forbearance, humility, love, loyalty to truth, to beauty, to goodness, and to holiness in a kingdom not of this "world." Life consists, not in making friends and fighting enemies, but in a fight with one's self to realize unselfish ideals, to exemplify the highest principles and laws, and to achieve the largest and best work, without regard to self-conservation or self-aggrandizement. In this radically new evolution the mind is for itself, and is not, as in the lower evolution, merely a utilitarian factor, subservient to the general demands of life. Life, on the contrary, here becomes subservient to the development of mentality purely for its own sake. Thus pure science, art for art's sake, an independent morality and religion, become possible. The greatest minds of the race are those who have lived most completely this highest life; but this new form scarcely touches the great bulk of humanity, and is very partially developed even in the so-called highest classes.

But it is not our present purpose to survey the higher evolution, or to point out its *rationale*. For the lower evolution, however, it is tolerably evident that fear, anger and hate, give the dominant tone to psychic life. These strong, direct emotions act as fundamental life factors; without them the individual would be quickly overwhelmed in the struggle for existence. The conditions of early life absolutely require these simple, naïve emotions to stimulate advantageous reactions. Emotional indifferentism is possible only as an artificial and by-product, a 162sort of disease or abnormal symptom even in the very latest phases of human evolution. The comparative psychology of the future will show more and more clearly and fully the nature and function of both the fear and anger groups as factors in biologic evolution.

CHAPTER XI

SURPRISE AND DISAPPOINTMENT, EMOTION OF NOVELTY

To anticipate what is to occur is plainly one of the most useful achievements of mind, for all providence implies apprehension and emotion therewith. But to look before and after is certainly not the prerogative of man alone, but anticipatory power is found throughout the realm of mind, and constitutes the larger portion of all cognition. To know a thing means, in general, to appreciate its potentiality; and all science is really prescience. Knowledge is not the immediate sensation, but the meaning of it for life; it is the ideal translation from one sense to another in feeling tendency. Thus, to scent is by itself a useless acquirement, but the connecting it with desired food is of the utmost service. The psychism gradually attains the power to interpret by various *media* the nature, that is, the experienceability, of the environment.

To foresee is then one of the commonest events in mind, and according to the painfulness or pleasurability foreseen is felt anger or fear, hope or desire, or allied emotions. But the foreseen does not always come to pass, and hence there results a new order of intellectual and emotional reaction. That what we had in mind would happen comes not, or is other than foreseen; this has a disturbing effect on cognition and emotion. Prescience defeated becomes 164not merely nescience, but there is the positive definite shock of surprise, and the emotion of disappointment or some correlated form. Surprise as the sense of contrast of real and ideal, involving personal sense of limitation and error, is, as we have noted (pp. 50 ff.), a painful experience. But where there is no preconceived notion, no expectation, there is no surprise, as Lumholtz remarks of the Australian savages, that they are not surprised at the railway and other wonders of civilization; they do not know enough to be surprised. The full apprehension and understanding of the gap between ideal and real is but very slowly attained. At first the thwarting is naturally and easily attributed to an enemy, and there is anger and pertinacious violence, but ultimately, by sad and repeated experience, mind is led to notice its own insufficiency, to feel that the conflict between the actual and the expected is due to subjective error rather than objective interference. Genuine surprise, as distinct from mere nervous shock, is then, I think, a later phenomenon than is generally supposed. What is often taken for surprise with animals and children is really eager attention. Again, certain modes of fright are often taken for surprise. But experience must have made a considerable advance in apprehension of experienceability before a real surprise can be manifested, which is always the correlative of a sudden contrariness of experience to what was preconceived. Surprise involves a certain measure of a theory of experience; in short, a more or less definite body of knowledge. One who has framed no ideas of what experience should be can never be really surprised at whatever may happen. However, to be able to feel surprise is obviously very advantageous, to have a painful and sharp sense of the incongruity of real and ideal often conducts to that investigation which results in being prepared against being surprised in the same way again. The imperfectness of adaptation is thus consciously and intelligently remedied. The man of 165large resources, cautious nature, and keen insight and foresight, is little liable to be surprised, for in all circumstances he accurately forecasts a very wide range of possibilities.

When the good expected comes in less measure than was foreseen, or not at all, or some real evil instead, there is not merely surprise, but disappointment as well. When what is confidently expected does not happen, the emotional reaction is surprise; when what is eagerly hoped for does not occur, disappointment is the result. I am disappointed in not receiving a certain remittance I had hoped for. Here the ought to be, the expected, is ranged over against the actual not, as in surprise, as a sudden and painful change in cognition, but solely for the personal advantage missed. Disappointment is bound up with the sense of personal loss and detriment from the happening contrary to expectation. Feeling of disappointment is thus emotional reaction from cognizance of evil result where good is looked for. The more it was hoped for, the more bitter the disappointment. This disappointment has its function as an emphatic protest against impracticality; the lessons of experience are thus brought home and made memorable. Disappointment turns life from false dreams to stern realities; it prompts to an investigation of causes, and rouses cognition to a full understanding of the situation. Hope thereby becomes more and more rational and realizable.

In all disappointment we note that the feeling is not about the past as such, but is with reference to the immediately actual in its unexpected bearing on life. Thus it is not strictly retrospective emotion. Though often initial to regret and grief, it should not be confounded with these.

A curiously illogical remark, and one not uncommonly heard, is, "I hope you will succeed, but do not be disappointed if you don't." This is really a psychological 166Hibernicism. Hope is the foundation of disappointment, and one cannot say, "hope, but do not be disappointed," in the same breath with definite meaning. We cannot escape the painful implications of unfulfilled desire: we cannot both have our cake and eat it too. Some measure of expectation of success is implied in all futuritive effort, hence a like measure of disappointment. The real sense of any such admonition can only be for moderating desire, and so tempering possible reaction. The expression in question amounts to little else than a phrase of well-wishing, but with little confidence in the actual result.

From the feeling of surprise and its congener, disappointment, it is natural to turn to the feeling for novelty. Surprise and novelty both relate, but in different ways, to the character of the experience in relation to other experiences. The strangeness, however, in what is surprising, and which makes it surprising, is not intrinsic, but wholly relative to a preconception. Thunder is familiar to me, but it may surprise me if it occur in January, and also totally out of my preconceived order; but a friend who has neither heard, nor heard of, thunder, will not be surprised by the sound in January, though he may be startled, and may feel the novelty of the phenomenon. The novel, purely as such, cannot surprise, for there is no field for the expectation which is the foundation of surprise. The surprising is always contrary to expectation, but the novel is simply unexpected, not in the range of thought and conception in any manner. A novel experience is one which has previously been unexperienced, and the feeling of novelty is the feeling of it as such, while a surprising experience goes quite against all we look for, and is often familiar and common enough, though sometimes it is novel, as when the absolutely new experience and not some familiar experience comes in place of the expected experience. If the man to whom thunder is novel is awaiting merely the pattering of rain, the crash of 167thunder will excite both feelings of surprise and novelty. In this case he is surprised before he feels the novelty of the surprising event.

A feeling for the novelty of an experience implies sense of experience and experienceable, and is thus debarred from primitive consciousness, which is merely a series of disconnected flashes, occurring a few times at the critical moments in an organism's life. It is probable that in the origin of mind the first consciousness was the last, an entirely unique and isolated phenomenon in the animal's life, hence supremely novel. However, at first, and undoubtedly also in later mind, consciousness but slowly rises to the sense of novelty of consciousness as such. After a long period of unconsciousness from any cause we do not appreciate returning consciousness as *per se* a comparatively novel phenomenon. In early mind every experience is practically a new experience, and so novel, but as there is no cognizance of experience in any light, and least of all in this light which is rather remote from immediate practicality, the feeling for novelty does not occur. Sense of novelty implies a comparison of experience purely for its own sake, certainly a very late acquirement. Thus in primitive mind, though all experiences are uniformly fresh, yet they are not appreciated as such. The feeling for novelty must always rest upon a considerable body of experience unified by ego-sense and apprehended as such, that is, consciousness of novelty implies both consciousness of consciousness and self-consciousness. The consciousness of novelty is thus far from being equivalent to novel consciousness. Whenever, even in advanced mind, a novel consciousness occurs, we should be over hasty if we at once concluded that feeling of novelty was also experienced.

The first step in life is to get an experience, to struggle into a consciousness which may be immediately valuable, and which is at once emotional and motor in its action; 168the second step is to compare and identify the experience gained so as to ascertain its meaning for life with greater certainty. Recognition thus comes early into play, but while the sphere of the sense of the novel lies in that of the unrecognised, it does not in any wise occupy the whole, for much that is unrecognised still is far from conveying feeling of novelty, because this feeling is, as we have said, far from being experienced on every presentation of the novel. The novel is equivalent rather to the unrecognisable. A dog may lose in a few months the power of recognising its master, yet the master after such a lapse of time cannot be said to awaken sense of novel. Though not recognised for master he is recognised as one of many familiar objects, he is known to be a man, and that is as far as the identification goes. The experience then is in reality not a fresh one. Here is a new man but there is nothing novel in the experience, much less is there a feeling of novelty. I doubt much if a dog or any lower animal notices and appreciates pleasurably or painfully the novel as such. The unrecognisable and unclassifiable presented to them may agitate them in various ways, as contrast a horse and a courageous dog on first seeing a locomotive, but there is no evidence of real feeling of the novelty of the experience as such. The enjoyment of the novel for its own sake is probably wholly confined to late human psychism.

It must, indeed, be granted that change from monotonous or confining circumstances is appreciated and appreciated pleasurably by lower animals, though they may not know enough to seek change for its own sake. Animals certainly suffer from *ennui*, and enjoy variety within certain limits, but change is not newness, and absolute change or novelty in strict sense hardly appeals to them, that is, they do not appreciate the novelty of a situation. The really novel disturbs them, they do not desire it nor are pleased with it. It is only in fact in the 169higher ranges of human mind that experience of any kind, novel or various, comes to be sought for its own sake. To say, "this is a novel sensation," or "how novel and delightful," and all similar expressions, denotes a frame of mind which is artificial, that is, lies away from and beyond the common course of psychism under natural selection. The changefulness of experience and the novelty of an experience are in reality two distinct elements. One who has been ill in bed for weeks enjoys the change in sitting up in his arm chair, but there is no real novelty or sense of novelty. Everything, we say, is novel and interesting to the child, tiresome and a bore to the blasé man of the world. The world is, in truth, fresh and new to the child, but the sense of the novel *per se* is very slowly developed, and the rarer the novel becomes, the more keen our appreciation of it. Where all is novel, there can be no sense of novelty, for this is purely a contrast type of psychosis. The zest and eagerness of the child proceeds from radically other sentiments than the feeling for novelty; it is absorbed in things for themselves and what they directly give, and does not stop to reflect and feel about the relations of experiences, and so feel the novel as such. Further we note that pleasing novelties are far from being equally pleasing as such. It may be as novel to carry a potato in my pocket as a double eagle, but not equally pleasing. The real value of novelty for emotion must always be carefully determined by subtracting accessory feelings.

With regard to the relation of novelty to pleasure and pain, the novel and the sense of the novel is always in its inception under evolution by natural selection unpleasant and painful. A novel experience is one which can only originate in painful struggle, and the new is always *per se* distasteful to early mind, which is ever conservative in its instincts and tendencies. A perfect life, biologically speaking, is one which is perfectly adapted to its environment, 170and so goes through its evolution with mechanically exact adjustment to circumstances; and the novel would break in upon the unconscious rhythm which is here perfected. Habituation becomes so iron fast that the novel, even when distinctly pleasurable in itself, is resented, much less is the novel sought for its own sake. However, so far as a novel experience may come rather by way of regressiveness than progressiveness, it may delight us by its novelty whenever the mind becomes capable of appreciating novelty. Thus purely hereditary tendencies, which we do not accomplish but which are accomplished in us during youth, as, for instance, the sexual evolution, may charm, not only in themselves, but for their novelty as well. But this experience which is not merely novel to the individual as springing up spontaneously by *impetus* from the past, but which is novel for the race, and requires effort to assimilate, and so is in the distinct line of higher evolution, as, the achieving a high spiritual sentimentality in love; this, the real novel, is inevitably and naturally painful. The first time the emotion of humility—a comparatively recent evolution—was experienced by a human being was a truly novel experience, though it is quite uncertain whether there was with it either sense or sentiment of novelty.

If the novel and the novel experience—and these terms are practically identical—are essentially painful, whence and how arises the peculiar pleasure which we undeniably may experience in connection with the novel appreciated as such? Must all such pleasure be placed to the account of regressiveness? But pleasure of this kind is intrinsic in the act itself and not for its novelty *per se*. There is a wide variety of experience intrinsically either pleasurable or painful, which may be pleasurable to us solely by reason of its novelty. I may enjoy the novel experience of tasting a pomegranate, be the actual experience agreeable or disagreeable, merely enjoying the novelty as such. 171What is this novelty, why is it noticed, and why does it give occasion to pleasure or pain in emotional form?

As we have already pointed out, the sense of the novel and emotion about it cannot be said to arise with novel experiences in general. The novel in the objective sense is the first occurrence of any given definite kind of psychosis, as humility or pity, in the history of mind, and this novelty is probably not at first appreciated.

Bain says that novelty is not an emotion, but "merely expresses the superior force of all stimulants on being first applied." But from the point of view of psychic history the initial force of stimulants is always very inferior and slight. For example, to taste and to qualitatively distinguish tastes is an extremely slow growth in the race, and by no means suddenly completed even in the offspring of the most advanced individuals. Place a drop of wormwood extract on an infant's tongue and it may have a novel sensation and a disagreeable one, as evidenced by the reaction, yet the real force of the sensation is certainly quite inferior to that of a ten year old child in the given case. The absolutely new impression is always slight, for mind is, in the natural course of evolution, always slow at fully experiencing things, it is by effort and by effort alone that it attains the several orders of sensation and perception, and it is only by effort that they are realized with greater and greater force and clearness. By the very nature of psychic evolution as a progressive process toward helping adjustability the novel exercises at the first but a slight reaction. However, in the exigencies of existence the most wide awake, those most susceptible to perceiving novelties and new circumstances and to being suitably affected by them, have the advantage. Hence the apprehension, interpretation, and application, of novelties is the path of progress which finally culminates in the achievements of human invention. An openness to the novel is thus of prime importance in 172a practical way, though this is quite distinct from the pleasing sense of novelty. However, the novel is not primarily attractive and interesting in and for itself, but this must be accounted a late evolution in an artificial period. The novel is at the first anything but charming. The absolutely novel is never pleasant for its own sake.

It is only in a relative way that the objectively novel pleases, that is, in the way of variety and change. Where overflowing mental energy by reason of habituation finds no full and easy diverse activity the mind is hampered and constrained. Thus youth in particular finds delight and relief in new sights and sounds, in fresh experiences of all kinds. Quickly wearied and exhausted in one channel and yet full of active power, the mind springs rapidly from object to object along those lines which ancestral experience has rendered the lines of least resistance, thus especially in the plays and sports of childhood.

While the novel in this way as change pleases, yet there is no pleasing sense of novelty. Sensations, sights, sounds, tastes, etc., please by their novelty, there is a pleasure in the sensations not merely intrinsic but relative to previous experiences, but the mind is not yet capable of the emotion of novelty which belongs to reflective consciousness. The child may be pleased by the novel, but is not consciously charmed by the novelty. The sense of experience as novel, and as such pleasing, belongs to a higher grade of consciousness than the naïve direct consciousness of the child. Novelty consciously known, appreciated, and sought for its own sake is a decidedly late evolution. There is an emotion and emotion of pleasure which we may feel in view of the novel *per se*. Not merely the new object becomes the stimulant of a new and refreshing experience, but this experience being known as novel by the reflecting consciousness, and contrasted with other experiences, there comes therewith a peculiar ripple of pleasurable emotion, the emotion of the novel. The first 173emotion of novelty is itself thereby a novel consciousness which might be, to a very reflective self-conscious mind, an object for another emotion of novelty. In touching upon the emotion of novelty we have thus risen beyond the common course of natural selection, to the point where experience values itself for its own sake.

In contrast to the emotion of novelty is the emotion of familiarity. This might be discussed in a strictly parallel way to our discussion of the emotion of novelty. It is founded upon likeness, being the sentiment of likeness. An absolute novelty, the perfectly new, is of course imperceptible as such, and by the law of continuity cannot occur in nature. Some correlation with past experience is required to make the thing cognizable at all, as is also some measure of unlikeness to make it distinguishable and so familiar. The emotion of familiarity is much neglected by psychologists, yet it forms a more important and a larger element in the pleasures of advanced mind than the emotion of novelty. Many of the delights of home and domestic life are tinged by it. The pleasing sense of familiarity is, of course, most felt in contrast after some long experience of novelties, as when the traveller returns home from a prolonged journey. Delight in the familiar for its own sake often largely prompts to the revisiting old scenes and renewing old habits. The emotions of novelty and familiarity have a constant contrasting play in many men. The familiar which is painful in itself may yet, like the novel painful in itself, be pleasurable. We often welcome the familiar and novel purely for their own sake whatever be their actual hedonalgic[C] content.

[Footnote C: This adjective, which I used before seeing Mr. Marshall's "algedonic," more exactly expresses pleasure—pain quality.]

Noticed familiarity like novelty may be painful. The disgusting emotion by which we may meet the unwelcome novelty, has its correlate in the wearing sense of monotony 174from the regular return of the familiar even though it be intrinsically pleasurable.

In the reflective emotions we have touched upon but a single group, the novelty-familiarity, which is certainly a complex but interesting kind of psychoses. In all this field we have rightly to separate mere sensitiveness to likeness and unlikeness—a tolerably early phenomenon—from sense of relatedness and unrelatedness of experiences in and for themselves. Consciousness of experience as such is the mark of a radically new type of consciousness, quite set off from the naïve unreflecting consciousness under the primitive conditions of natural selection and the struggle for existence. The significance of this, by which experience rests purely upon itself and is for itself, leads into a wide region. It is enough that we have instanced one of these later emotions in contrast to the directly serviceable emotions which have most concerned us in our present discussions, without inquiring closely into its function. It is evident that in the ordinary course of evolution the character of the situation as affecting life determines the serviceable emotion, thus different kinds of harmful situations determine fear, anger, hate, etc. If a situation is really interesting for life, it ultimately will be both known and felt in the progress of the struggle for existence just as surely as light, colour, sound, etc., are gradually appreciated. Hence we might predict that the novel situation and the incongruous situation would receive some advantageous cognitive and feeling response, and that even emotions of novelty, familiarity, congruity, and incongruity, would arise, as well as the feelings for these things, if this were useful; that is, experience may ultimately consciously react upon itself in these ways as well as directly sense mere objects. Now the pleasure in novelty for its own sake, while not consciously in the region of natural selection, yet indirectly may be favoured by it as propædeutic to progressiveness. It would, indeed, from one 175standpoint seem possible to deduce according to the law of serviceability the whole course of experience past, present and future, and we might as assuredly predict particular feelings as we may predict the evolution of the wing or the hoof or the four-ventricled heart in the course of a physical biologic evolution. The psychic biologic evolution is to a certain point as strictly interpretable by the principle of advantageous natural selection as the physical, for the two are really co-ordinated. In the near future of psychology every psychosis in its origin and development will be as clearly traceable as any purely physiological organ, though this can never be accomplished in the purely objective manner, but will require a subjective manipulation which is now quite beyond us.

CHAPTER XII

RETROSPECTIVE EMOTION

Brown divided emotions into retrospective and prospective, but such a classification has no basis in a general biological view nor yet in a special analysis of the particular phenomena. It is evident that the two great classes of emotion from the point of view of struggle for existence will be response to injurer and to benefactor. These are the two prime qualities in things for which emotional notice is most needed as a service to life, and hence the broad and fundamental division of emotion must always be into that which is response to the harmful and that which is response to the beneficial. Here only is the great and constant distinction in the nature of emotions. Prospect and retrospect are equally meaningless in themselves considered. From a merely *a priori* biologic point of view we must, then, pronounce it quite unlikely that the time-sense should fundamentally differentiate emotion, but we should expect that the prime division would be with respect to cognised injury or benefit.

That time-sense is not a grand principle of division we also see plainly when we examine particular emotions. Thus, in the case of anger, while we can say at once that this is, in all its forms, repulse to injury, can we claim it is either prospective or retrospective emotion? The truth is, the thought of injury done, doing, or to be done, equally wakens anger in choleric individuals. The man who harmed me yesterday excites my anger, and so does the 177man whom I perceive to be now injuring me or about to injure me. The quality of the emotion is identically the same whether the object be considered as in past, present, or future. Even what seems to be a purely temporal emotion, like hope, which is usually regarded as wholly prospective, may yet have other temporal aspects. Thus, we sometimes say, "I hope it was not so," where hope is obviously retrospective, or more strictly prospective-retrospective, having reference to expectation with desire that the event will turn out not to have happened.

But it may be said that, as emotion rests upon representation, the proper classification of the emotions will depend upon the divisions of representation which are essentially determined by the time-sense as representation of past or future. Representation with sense of representation implies a cognition of the thing as represented merely, and so as non-existent to present actual sensing, as something having been, or to be, sensed. The emotion arises thus on cognition of the experienceable, and includes always some dim impression of potency of object for harm or benefit at some time. However, though this may be the case, it is plain that it makes no radical distinction in emotion. If a man threatens me with some injury, this fires my rage, which is greatly increased if I catch him in the act of committing the injury threatened, or find that he has committed the evil deed. Change in time-sense may thus bring change in intensity of some emotions, but it does not determine quality of emotion. The prime factor as to kind of emotion is always, not any sense of time, but the personal value of the event, which may or may not receive a definite time determination. Indeed, a form of representation, before any sense of experience as merely subjective phenomenon is attained, is a prominent feature in the direct naïve experience which constitutes by far the greater bulk in the total existent consciousness. Before experience is aware of itself and of the experienceable 178there is a certain purely subjective mirroring of that which is not present to sense, but has been, *i.e.*, there is a re-occurrence in consciousness which has the subjective force of reality; though the objective actuality is lacking, such re-occurrence by association without the actual presence of the object stands, however, for reality to the mind experiencing—it is a direct intuition; the object, though unreal, is perfectly real to consciousness, and conveys no meaning, and so is not a basis for emotion. Yet in the higher representation with a sense of experience as integral element, the representation is sometimes practically timeless, though surcharged with emotion tendency. The highest objects which the mind represents have little time quality, and all the nobler sentiments, as love of truth, justice, etc., exist with little or no reference to time. So also in the very earliest representation, the object is seen in its feeling value—emotion basis—as soon as it is perceived as object; but this is as an immediate subjective realizing in which time-sense plays very little part. The conscious interpretation of past and future as a conscious connecting of the two is certainly not a primitive function. The time form is, then, on the whole, merely incidental in emotion, and is by no means a fundamental principle determining classification.

Yet, though we must reject time as a cardinal principle of division in emotion, still we must acknowledge that the term retrospective emotion denotes a real group of mental phenomena, including revenge, regret, remorse, and kindred forms, which are marked as feeling for the past merely as past. However, pure retrospection is rare and late. The past does not for primitive mind stand by itself as something to be dwelt upon, to be thought about, to be moved by, and stirred to action. The immediate present absorbs the mind, and the past interests and excites only so far as bearing directly on the present. And so it is that the child lives in the present, the youth and man in the future, 179the old man in the past; and this denotes the relatively late appearance of pure retrospection and of emotion founded thereon. Emotion is first merely spectant, then prospective, then retrospective. However, when we say an emotion is concerned solely with the present in the very young, we mean, of course, the immediately prospective—that which has relation to but one sense and by association rouses emotion, as an apple, seen or handled by a child, awakens emotion, desire to taste. Where sense consciousness is not multiform, but single and uniform, as, doubtless, in very low organisms, there is no opportunity for any emotion, for there is no interpretation power. But the intensification of some one sense connection already attained may be a basis for emotion which we may loosely call emotion spectant, as when the greedy child eagerly eating an apple desires a larger bite, sweeter portion, etc. However,—though it has little classification value,—emotion can be only prospective or retrospective; and this is, of course, implied in its basis—representation. Emotion by its very nature must be a looking forward, or a looking backward, or both. As a feeling about, and not a direct feeling, this is obviously its unvariable cognitive content. The immediate and actual realization may be direct feeling or sensation, but it is never in itself emotion. Emotion is always over something, an experience of experience, and cannot thus be simple content. It is thus a consciously idealizing mode as distinguished from direct realization which is wholly self-contained.

One of the most important and interesting retrospective emotions is revenge. The cardinal idea in revenge is returning evil for evil. Not only must there be a paying back for past injury, but there must be an equivalence, eye for an eye, and tooth for a tooth; and the revengeful emotion is the meting out such purely retributive action. Exact return becomes the basis of a general usage in animal and human societies. Justice, law, and punishment 180rest upon the idea of inflicting duplicate or equivalent injury for injury received. Administrative justice is the specialization of revenge in the hands of a few members of a community, a social differentiation by which individuals in general secure their revenges at great economy by proxy. Further, the revengeful emotion is a smouldering hate which vents itself only some time after the immediate occasion. This is not the flush of anger which prompts to vigorous offensive action upon the injurer at the very moment of harm perceived, and it does not appear as stimulant to immediate self-conservative activities, but is simply the spirit of getting even for relatively long past injury.

What, now, is the function of revenge as a life factor? It surely does not mend my injury that I do another harm solely because he has some time harmed me, and the whole impulse might seem a pure waste of energy. But under natural selection revenge must arise in serviceability of some sort; and it is obvious that while revenge is of no use in mending the past, it yet has a large value with reference to future possible injury. Yet revenge is undeniably without conscious meaning for present or future; it is merely the spirit and determination to get even, and so its deterrent function is unconsciously attained. A dwelling in thought on the past *per se*, a feeling about it and acting on it, while it cannot help life directly, has a large value in its ultimate effect upon enemies. He who never forgets injury, and for whom by-gones are never by-gones, who never fails to return injury for injury, is feared and is less likely to be injured. Junker, the African traveller, remarks of the pygmies, "They are much feared for their revengeful spirit." Thus, other things being equal, the most revengeful are the most successful in the struggle for self-conservation and self-furtherance. Though by itself considered irrational and foolish to inflict return injuries upon an injurer long after the immediate occasion, yet its 181deterrent effect is very great with reference to other assailants. Thus, pure retrospection may have unconsciously prospective value, or sometimes revenge may be really retrospective-prospective, as when one says, "I will fix him so he will not do that again." Here function is consciously known, but in instinctive revenge there is no such foresight, and, in general, utility is no consideration with the revenger, whose mind is bent rather on doing great harm for its own sake to his enemy rather than benefiting himself. It is always the conscious or unconscious significance for the future that justifies revenge in the natural course of events; while it is no remedy for my hurt, if some one has put out my eye, to put out his in return, yet this revenge act, and so the feeling which prompts it, is of highest prospective value with reference to future possible enemies. Every one will know that I cannot be harmed with impunity. Despoil or injure the revengeful in any way and you inevitably suffer for it sooner or later, and so revenge acts as a protective psychical variation of high value. On the whole the revengeful is less likely than others to be molested and injured, and thus has a manifest advantage in the struggle for existence. Revenge has, then, also rightfully its own subjective sanction, a pleasure reaction, for revenge is, indeed, "sweet."

Revenge is apparently found in a considerable range in the animal kingdom, and seems universal in the *genus homo*. However, we cannot infallibly conclude from certain actions that revengeful emotion is present, and especially is this so in the case of animals. Thus, in the well-known instance of the elephant, who, observing a man passing by who had greatly annoyed him years before, suddenly drenched him with dirty water, we are not necessarily to suppose that this elephant was prompted by the emotion of revenge; although this may have been the case, we are not perfectly sure how far the elephant did the act merely as recompense for what the man had done, or how far the 182sight of the injurer, and so one likely to injure, roused to simple anger and defence against the threatening harmful. Many acts which seem like revenge are quite likely to be common defence or offence, are done with reference to what the object is and will be as injurious, based upon knowledge of the past, and not as merely retrospective retributive acts. Memory for injuries received is strong in many animals; that which has harmed is often recognised after many years as the harmful, and appropriate simple emotion, not revenge, is manifested. Rage, rather than revenge, is the usual emotion among lower animals in special instances where revenge might seem called for; and thus it is more likely that the elephant should rage and hate rather than have pure revenge as in the case considered.

However, somewhere rather late in sub-human psychism revengeful emotion certainly arose as an advantageous variation, and it grew in strength and prominence for many ages of psychic progress. At length it culminated, and began its decline with the marked increase of co-operative sociality, with which it must greatly interfere. Reprisal and counter-reprisal, vendetta, feud, is opposed to that social union which is strength; and so we see that tribes and nations in which the spirit of personal revenge has been a dominant trait have been left behind in the march of progress. Revengefulness, at least in the form of retributive personal violence for injuries done, is, in a highly civilized community, entirely superseded by the machinery of law. Instead of slaying a brother's murderer I call upon the law to execute justice and retribution, and I bring certain designated ones among my fellows to secure my revenge. Where a man takes the law in his own hands, and kills or injures the violator of his home or the slayer of his nearest kin, he recedes to the lower unsocial plane from which civilization has arisen. Thus revengefulness, in certain forms at least, has become in the highest human 183communities a disadvantageous variation, and is gradually being eliminated. This negative elimination of revenge is also greatly hastened by the progress of certain ethical and Christian conceptions by which a new and opposite law of conduct is enforced, namely, the returning good for evil.

One of the most interesting and most retrospective of emotions is sorrow. Sorrow, grief and regret are wholly regardful of the past, are pains at the past. They are purely subjective or "mental" pains at the past, and in no wise pains from the past; they are not pains recurrent from past pains, but purely a painful emotion at the representation of past pain. Thus, a man says, "I did it to my own harm and hurt, and I have always been sorry I did it." Here the sorrow-pain is evidently quite distinct from the direct pain of the injury; pain for the harm done is one thing, and pain from the harm done is another. I hurt myself, and I not only have this pain, but, being sorry that I did it, I have this new emotional pain added. Sorrow as painful emotion for the past is thus plainly unique and peculiar. To feel sorry over what has happened is a mode of feeling altogether different from feeling proud of it, angry at it, etc., and we may reasonably regard sorrow as a distinct *genus* of retrospective emotion. What, now, is the nature and function of this special emotion reaction?

We have to consider here only that simple primitive sorrow which is a painful emotion at regarding personal loss or failure. Such simple sorrow we see in the child who cries over spilled milk, in the man who expresses deep regret at the careless misstep by which he broke his leg. In this emotional reaction at the injurious the harmful is neither escaped nor repelled, as through fear and anger; the feeling disturbance is comparatively passive and purely reflective, and is not a spur to some immediate advantageous defensive or offensive activity. In sorrow we are pained emotionally at the trouble which has come upon 184us through our own agency or otherwise, but we do not struggle from it or against it, but there is purely helpless retrospection. Harm and loss which might provoke in one nature to fear or anger, in another lead only to inactive sorrow.

The cognition form in sorrow means always sense of *personal* loss. I may fear a thing, or I may be angry at a thing, but I can be sorry only for a person. I do not feel sorry for a broken chair, though I may feel sorry for having broken it. This view of one's own personal agency in causing harm to one's self and harm to others is very prominent in a large range of sorrow. In viewing any action which determined some evil, I say, "I am sorry I did it." This is, however, a later mode of the emotion, which at the first cannot take account of any agency, but is simply an acute feeling of distress at the injury received. Thus the one who grieves over the spilled milk regards, not his own agency, but only his loss; he is sorry, not that he spilled the milk, but that his milk was spilled. Yet the sense of personal agency certainly forms a great part in much sorrow, and tends to intensify it. I may grieve over any harm that has come upon me, but my grief is intensified as I remember my own agency in bringing it about. I may feel sorry over the loss of my goods by fire, but if I lose them by my own careless act, my sorrow is redoubled. Strictly speaking, perhaps, the sorrows are distinct, I feel sorry for having done it and I am sorry at it done; yet they may be said to constitute a single psychic state. Sense of our own agency, however, in having produced harm to self is as likely to produce anger at self or even fear of self. Hence our intensest and purest sorrows are apt to be those occasioned by considering injuries occasioned by elemental forces. That harm which we did not help because we could not, the inevitable injury, this excites a keen regret and deep mourning.

185The pain in sorrow is as peculiar, searching, unanalyzable and undescribable as other simple emotion pains, and only conceivable through realization. This sinking, helpless pain over what has happened is clearly distinct from the sensation order of pains, and is in no wise a reflection from them. The pain I have at remembrance of some great loss which has befallen me is certainly very distinct from that which came from the loss itself.

What part now does sorrow play as a psychic life-function, and how explain it on the general principle of natural selection? At first sight, sorrow or grief over the past seems utterly valueless, seems to be mental energy thrown away. The past is irretrievable, of what use then is any grief? Is not all regret vain? To deplore its loss does not tend to restore a lost arm, and it is of no use crying over spilled milk. Indeed, he who bewails spilled milk has not only the actual loss but the ideal pain about the loss. He who grieves suffers doubly. But while it is true that sorrow for what has happened cannot alter the occurrence, yet it has a permanent salutary effect on the one who sorrows to give more caution for the future. The child will carry the pitcher of milk the more carefully next time by the more he has grieved over the past occurrence. By increasing sensitiveness and capacity for sorrow experience is strengthened, deepened, and completely adjusted to environment. Shallow and volatile natures, who take all loss and harm easily, and even gaily, have little strength, and attain no great and permanent growth. But with most, when the object of strong desire is suddenly lost, not only will there be a disappearance of the positive feeling about it, but an actual *minus* or negative state will be generated, a reaction mode we term grief. By this grief the chief lessons of all higher experience are made possible. Grief is not a pathological phenomenon in mind, but in its place thoroughly normal and useful. Indeed, if under certain circumstances grief did not appear, mind 186would be proved very crude, obtuse, or diseased. He who never feels sad about what has happened, is not of a progressive or highly advanced type. If one does not feel sorry for his past errors and hurtful actions, he plainly has so much the less motive force to higher action for the future. If sorrow had never entered the world of mind, if the whole corrective for injurious actions or want of action lay wholly in the immediate pain resulting or in the direct simple emotions like fear and anger, a most potent factor in psychic progress would be lacking. The possibility of going wrong, *i.e.*, literally aside, and contrariwise to one's own interests, is implied in the struggle for existence. The next best thing to the impossible *status* of being unable to do wrong, is to have the capacity of feeling for the wrong, that is, of experiencing grief. Sorrow is thus a corrective of the highest importance in the history of experience. The slips, willed and unwilled, from the narrow path of upward evolution are of necessity many; but a man is, on the whole, best doing the largest part in the evolution scheme in which he finds himself, who both knows the wrong as such, and is sorry for it, whether in the primitive selfish mode, or better still, on the higher ethical and religious grounds. The greatest and most efficient minds are those who have felt most keenly for their errors, faults, and sins.

As to the origin of grief, we may say with confidence that it is tolerably late, and certainly subsequent to anger and hate and like reactions. Under certain circumstances sorrow must be accounted a more favourable reaction than these. Rage is certainly impotent and useless on many occasions of recalled injury, and rage is besides a very intense emotion and expensive of energy. The general law in the development of emotion is toward milder, more economical, and more permanent forms, and then it is that sorrow must at some time have originated under the demands of life, and been preserved and developed under 187natural selection. Sorrow most probably originated as supplanting rage at the view or remembrance of injury done. In young children we often see rage mingled with the first manifestation of grief, and but slowly is the rage eliminated and pure grief attained. Sorrow exercises its function where rage is useless. The child cries over spilled milk partly from rage, partly from grief, but such mishaps will tend more and more to be attended by grief only, as the better and more economical reaction. Further, in a certain range of cases, sorrow in its manifestations serves to appease revenger, and sincere regret, unmistakably expressed, often saves the wrong-doer an equivalent harm. This form of sorrow function is distinctly cultivated in the education of children where they are taught to feel sorry for faults if they would be forgiven and escape punishment.

Grief in its origin and its earlier occurrence is not the spontaneous and almost irresistible impulse of our adult human experience, but, like all emotion and all progressive psychism, is by effort of will. That is, we must suppose that grief has its origin in some such *nisus* as a child exhibits when he is taught to be sorry for something he has done. Hence it is only gradually and with the lapse of many generations after its origin that sorrow becomes hereditary and spontaneous. At first sorrow was a distinct attainment, rarely and but occasionally reached by any individual, and it is comparatively late in psychic history that it becomes a permanent and innate power. Sorrow also very gradually widens its sphere. At first purely selfish, a retrospective reaction at one's own hurt, it becomes at length, through sociality and its concurrent advantages, altruistic; sorrow is felt for others and the springs of sympathy and pity are developed. That this altruism is very late development is obvious, in that it has still to be taught even among the most advanced of the human race to their children. The child is taught to 188feel sorry for the cat he has hurt, for the blind man, for the cripple. And we must conclude that at one time in psychic history egoistic sorrow was likewise at the stage of development at which we now see altruistic, and we may suppose that in the far future the altruistic may come to the present *status* of the egoistic sorrow. However, for both there is an indefinite field for expansion, for refinement of sensibility, and for readiness and appropriateness of manifestation. Sorrow also will develop more and more on ethical and religious grounds. Remorse arises and develops; and also the "godly sorrow for sin." We learn to feel, not merely sorry over the past as affecting our disadvantage, but to feel sorry conscientiously as our deeds or those of others conflict with the law of right or with the law of God. Those who have no God-consciousness, and so no feeling about their action in the sight of God, no sense of sinfulness, have yet often acute moral sense and feelings. However, the origin and function of the moral and religious sense in the light of natural selection is a wide subject which can only be alluded to here; suffice it to say that sorrow is thereby lifted to a peculiar and new plane of self-contained spirituality. That is, the bearing of it is often without relation to physical life-function, and even adverse thereto, and throughout has its value and sanction in itself alone.

One of the deepest and most significant of late forms of sorrow is that for the dead, and its importance is obvious from the fact that a word is especially coined to denote its expression, namely, mourning. Nothing can be more useless than mourning for the dead as far as the individual object is concerned; the most poignant sorrow cannot in anywise tend to reanimate the corpse. However, it plainly serves as an index to the value put upon life, and so in general has a most powerful effect on conservation and upbuilding of life. Other things being equal, sensitiveness to this form of sorrow measures accurately possibly self-conservative 189effort or effort for others' conservation, which in a state of sociality, is equivalent in value to one's self. The lives for which there is the most mourning and real sorrow when death comes are the most valuable to the community, and for the conserving of which the utmost combined effort would be extended. Where life has little value attached to it, sorrow is slight and mourning short. As compared with the savage state, loss and injury to life is infinitely more respected in the great centres of modern civilization—the *nuclei* of progress. It is because we feel strongly for the safety of friends and relatives that we employ the best devices to insure their protection from injury and death. One who has sorrowed most deeply over the death of a friend caused by his own careless handling of a gun, will for the future be much more careful for himself and others. To be sure we sorrow deeply because we place a high estimate upon the life rather than place high estimate because we sorrow greatly; but if there were no sorrow reaction, there would be no emotion basis for the future caution and care, and it affects our general estimate of life. Thus there is ever a cumulative emotional development.

Perhaps the latest developed form of sorrow is the feeling of sadness which comes over one in reflecting upon pain as a universal fact of existence. The pessimistic mood, with its converse, the optimistic, as based on philosophic generalization, is certainly extremely late. Pain at pain in general, pleasure at pleasure as a purely general fact, are equally remote from primitive modes, and mark culminating phases. While, perhaps, there is a certain justification and value in being saddened by the spectacle of universal pain, yet a gravity rather than a despondency is its proper measure. Pain, punitive and premonitory, plays, as we have already noted more than once in our discussions, a most beneficent and essential part in the struggle for existence and in all the higher struggle. It 190is a necessary and salutary phenomenon, involved in the very nature of evolution by struggle; hence he who impugns pain and is offended at it, really impugns the psychic nature of things and desires with Schopenhauer the annihilation of will. As a matter of fact the extreme pessimistic spirit is more destructive to progress than even the most buoyant optimism, in that it nips all earnest and forceful activity in the bud. A foolishly happy-go-lucky activity is better than a paralysis of effort through conviction of its inherent painfulness and ultimate inutility. The scientific evidence, so far as we can now read it, points decisively to the belief that pain-will activity, an intense struggle, is the earliest mind, and the condition of its birth has been the law of its development, and for aught that we can see, ever will be. Into this we are born, and it is as foolish to run counter to it as to the law of gravitation. A philosophy which runs counter to reality must either build a new reality or subside; but it is most certainly to be doubted whether the philosophic spirit ever has or ever will determine a general innovation in psychic evolution. But we cannot do more than merely advert to these large questions here.

With reference to the development of sorrow it is an obvious remark that much which causes grief in the earlier stages of mental growth ceases to have that effect with maturer experience. Thus the man may not notice, or may laugh at, or may feel irritation at occasions which in his early life would have wakened grief. On the contrary, much that seems grievous to the old is not so regarded by the young. In general, grief tends to become less frequent and paroxysmal, but more profound and lasting with the growth of mind.

As to the kinds of retrospective emotion the largest division is, of course, into the painful and pleasurable. We have touched only on some of the painful, but each painful emotion has its analogous pleasurable emotion. 191We have used the terms sorrow and grief as synonyms. If we should make a distinction, it would be to put sadness or sorrow in antithesis to happiness, and grief to joy; that is, sorrow proceeds from outward circumstances, grief from subjective conditions. However, popular usage is not firm on this point. Regret is a mild sorrow. Remorse is the ethical side of sorrow. Resignation is a very late phase of emotion related to sorrow. A person says, My child was crushed in the accident, yet I do not grieve, but am quite resigned. Here certainly is a new mode of feeling about past harm, and it is a mode as far above sorrow proper as sorrow is above anger in the evolutionary scale. We do not lament or weep over the past, but there is self-conscious, self-constrained sinking of the will, and a composure which is not apathy, but a gentle emotion wave. Nor is there a callousness; one is not hardened, but softened, and made the more sensitive. The emotion of resignation is thus cultivated and to be cultivated, and is yet in the volition stage which marks the early form of all emotions. Even in the highest human types resignation does not come, it must be brought; the instinctive impulse upon contemplating past personal evil is toward sorrow or anger and revenge, which must be checked, and resignation directly willed and assumed as the proper emotion. Resignation, then, as a growing point in psychic evolution, a distinct attainment as frame of mind, is generally and rightly accounted a virtue. At present, then, it seems the culmination of retrospective emotion with regard to past personal injuries, and it exercises and will more and more exercise a most important function in human psychic development.

CHAPTER XIII

DESIRE

The lowest organisms come in contact with things, have objective relations of contact, but it is quite unlikely that the earliest psychic life feels contacts, really touches things. From the objective commerce with things pleasures and pains are realized, but object is unsensed and unknown. The simplest marine forms are incessantly feeding at hazard at the prompting of a subjective lack-pain. That the lowest life is born into a nutritive medium and that at birth many later organisms are incased or in direct connection with nutritive material, shows that at the very beginning psychic life is not needed as discriminatory, but as simple subjective pain and pleasure moving to undirected activities. However, such perfect environment being rare and temporary, in its blind and senseless activity the organism is often trying to assimilate the unassimilable, or the harmful, and is often appropriating when there is no substance present. It would obviously be of great advantage if it could touch its food, have sensation as guide to activity. Thus realization of a very limited world of things arises in touch achieved during the feeding act. That which satisfies and gives pleasure is by touch discriminated from that which does not give these results. Discrimination of soft and hard is probably the earliest touch impression. The soft thing is manipulated in the feeding act as edible. But a great step is made when psychical effect of the edible is not 193only comprehended through touch in direct connection with the assimilatory act, but antecedently thereto. The animal establishes a connection between the feeling the soft thing and pleasure experience in its struggling activities. It touches more and more readily what it is assimilating, and thence rejects more easily and promptly the injurious. In appropriative effort with pleasure experience it feels the thing, cognizes in most general way its physical quality. As sensitiveness increases through struggle and natural selection the assimilatory attempt will be more and more quickly met by the touch sensation, until touch ultimately becomes precedent and actually directive to food. Recognition, in a far more emphatic way than before, becomes added to cognition; the thing is not merely known in its bare objectivity, but is recognised, identified, and has a meaning. Touch must give, not only the thing, but the thing as potent for some quality not now being appreciated, though formerly appreciated *pari passu* with the touching. The interpretative act comes through the association gradually established in past experiences, so that the edible is no longer fortuitously hit upon, but touch precedes active effort at appropriation, and suggests by itself edibility or non-edibility. Thus is action greatly economized and made certain. Definite feelers, extending from the body, and sometimes quite long, are evolved, and the first period in the history of knowledge, the age of touch, is inaugurated.

It is here when touch involves representation and becomes a sign of something, *e.g.*, edible thing, that desire and other simple emotions originate. A possibility of pleasurable experience being recognised, it is necessary, if useful action would follow, that emotion springs up as incentive, and this emotion we term desire. Hunger drives, but desire draws, and as reinforcement and guide to the blind hunger impulse desire has a large function. A mere indifferent recognition, the pleasurable foreseen 194but not felt about, would be entirely unserviceable. If we do not desire the pleasurable and beneficial, we do not act for it. And originally, at least, perception of the good always stirred desire; and desire was awakened in no other way; for in the course of natural evolution, knowledge and emotions have alike to be interpreted in their origin and meaning with reference to advantageous action, this alone being the arena of natural selection. A meaningless knowledge and a self-contained emotion or feeling, are entirely contrary to the trend of evolution on the basis we have assumed. Moreover, through ages of activity the tendency to desire the good and the good only becomes so ingrained that I think it hardly fails, even in the highest and latest minds. The most hyper-conscious man, once convinced that something will give him pleasant experience, *so long* and *so far* as this feeling is *dominant* in mind will have incipient desire.

On this long disputed question of the relation of desire to the good or pleasurable, evolutionary psychology, which views mind as serving life, as interpreting things with reference to their serviceability and so implied pleasurability, always bases desire in its origin and growth on pleasure. But is this general point of view borne out by the facts of mind? A typical example of common desire is this: At a fair I observe a toboggan chute and say to my companion, "That must be sport, how would you like to try it?" The appeal to "sport" awakens desire in my comrade and he says, "Let's try it." We test its pleasurability, and, enjoying it, desire to go again. It is evident that desire arises not on the mere image of actualization as such, the idea of sliding, but on conception of its pleasure quality. Whenever by our own experience or by the testimony of others we are assured of a good thing to be experienced we straightway desire it.

This, it may be said, is all very true for a certain class of desires, but the principle does not apply in the higher 195desires like the desire for knowledge. But knowledge originates only as serviceable, and primarily only serviceable knowledges are desired. We desire knowledge only so far as it is worth having, and it may be that I esteem all knowledge as worth something and so desirable. However, some knowledges are worth nothing and are never desired. Who wants to know the exact measurements of the pebbles on the road, or how many hairs are on the mane of his neighbour's pony, or the names of all the inhabitants of Pekin? But if one thinks it would be any satisfaction to know such facts, he may desire to know them. The insatiable curiosity of children which seeks to know all such irrelevant facts hardly comes under the category of desire, but is rather instinctive hereditary impulse. It has no clear idea of a thing to be known and a desire to know it, but is only a spontaneous outburst of knowing activity which is inbred and comes from ancestral integration. There is a sensing and perceiving activity which is very intense at the questioning age, but which hardly implies the desire to know. The incessant "What's this?" "What's that?" is merely outcome of an instinctive impulse to interpret environment; it is not significant of full-formed desire, there is no idea of thing to be known, of an actualization to be accomplished.

If a man desires knowledge, not for his own sake, but for its own sake, desire as such really ceases, it merges into love and devotion, which are disinterested and clearly distinct as mental modes from desire. Desire is not a sentiment; and it does not properly include all impulse to actualization. For instance, the feeling for actualization merely as such, for achievement of ideal *per se*, is beyond the biologic stage of consciousness wherein desire has its chief function. The attainment of end merely for the sake of the end must be distinguished from actualizing an image for the pleasure of actualization, which thus has desire element. We know that the image of realization 196may act as end by compulsion, as in feeling of duty, which is thus marked off from desire as impulsion. Thus desire is but one mode of teleological emotion. But desire is emotion at unrealized good and not at unrealization in general.

Spinoza's *dictum*, followed by Volkmann, that we do not desire a thing because we deem it good, but we deem it good because we desire it, is not borne out by the commonest facts. A peddler shows me an apple, but I do not desire it and then deem it good, but I examine it, and if it seems good I may desire and buy it, but if bad, I have aversion, and return it. My desire thus depends altogether upon whether or not I deem the apple good, and not my deeming it good upon my desire. If I see any one desiring anything I at once judge that he first thought it good or he would not have desired it. All the excitation of desire is by representation of the good. The merchant tempts you by exhibiting his *goods*, the child with candy offers it to you crying, "good! good!" the moralist proclaims, "do this and thou shalt live." The cause of desire, which for weal or woe plays such a large part in almost all psychism, is always by imaging the good. The bait and the reward as excitants of desire are most common; a mere suggestion of a representation without implication of its goodliness in realization does not excite desire. Thus some one, speaking of a totally unknown town, asks, "How would you like to live in Perry?" and we answer, "Is it a pleasant town?" A mere suggestion of change of abode starts desire only when there is already displeasure with present residence, and so desire for release as a good; but image of actualization considered solely by itself is desireless. And if to excite desire we offer the good or pleasurable, to extinguish desire we offer the bad and painful. I desire a fair looking apple, but cutting it and finding it wormy and rotten, desire flees. I extinguish the desire of a child for 197eating some noxious substance by assuring it of the bad taste and nauseating effect. Both positively and negatively then, common sense finds the basis, not of the good in desire, but of desire in the good. The facts in both exciting and extinguishing desire point to this conclusion.

Spinoza (*Ethics* iii., Prop ix.) defines desire as "appetite with consciousness thereof." But to be aware of being hungry is but the first step toward desire. In the midst of my daily occupations I become aware of pain, then of uneasiness, then of hunger, whereupon I may desire food, which desire includes as distinct elements: (1) idea of eating as act or movement; (2) idea of the thing eaten as *food*, a something satisfying; affording relief, and so a good; (3) thereupon the emotion wave of longing, the essential point in desire. This is, of course, followed by volition, I act to realize, I go to a restaurant. When Höffding (*Psychology*, p. 323) says that the impulse in hunger "has reference primarily to the food, not to the feeling of pleasure in its consumption," he forgets that "food" is a something satisfying, and only thus is desired. Object is not desired as object, but for its value in experience.

We must also touch upon a certain class of experiences which have been adduced as showing a desire not based upon the idea of the pleasure. Take the example of a man in *ennui* who takes to playing tennis as a relief, but with no desire of being victorious. Engaging in the game he finds that "this desire which does not exist at first is stimulated to considerable intensity by the competition itself; and in proportion as it is thus stimulated both the mere contest becomes more pleasurable, and the victory, which was originally indifferent, comes to afford a keen enjoyment." (Sidgwick, *Methods of Ethics*, p. 46.) But does the desire really come from some idea of pleasure? The player volleys a ball successfully against his opponent, and thereby receiving a thrill of pleasure desire awakes 198to beat. "Wouldn't I like to beat him? I would enjoy nothing better." This desire foresees the pleasure of triumph. If he gets no pleasure from returning the ball successfully he does not desire success; but if unanticipated pleasure comes up in beating his opponent, as soon as he recognises this pleasure he desires to continue and complete it. This pleasure in succeeding in competitive activity, extremely old and integrated from all the struggle of existence, springs up spontaneously. There may also be added pleasure from activity and pleasure from skill which will make the game very interesting, *i.e.*, full of desire and other emotions.

Professor Sidgwick allows that pleasure may be the cause of desire, but not its object. But surely if I cognize pleasure coming from an act, I attach this pleasure to it in representation; if I take pleasure from returning a tennis ball and then represent a coming opportunity to return the ball I also represent its pleasurability. Pleasure or pain connected with acts is connected by association with representation of the acts, the pleasure-pain tone penetrates the representation, and only thus does actualization of an image become object of desire. If it is possible to conceive an activity indifferent—which may be doubted—we should have no emotion about it. But we have already sufficiently emphasized how the perceived experience quality of things determines desire and all emotion.

Professor Sidgwick's remark that the pleasurableness of the contest is "in proportion" (*Ibid.*, p. 46) to the desire, *i.e.*, that the pleasure results from the desire rather than desire from the pleasure, also shows defective analysis. If I desire intensely to beat, and am on the losing side, I am greatly pained, for desire is always in itself painful. In any case desire is pleasurable only so far as it is being satisfied, which, of course, means only so far as desire is being extinguished. It is not the increasing desire intensity, 199but the decreasing, that gives pleasure, *i.e.*, desire is negatively related to pleasure. Intense desire may act as excitement-pleasure, but this does not bear on the nature of desire.

Another objection that has been brought up against pleasure as desire basis is that "pleasures are diminished by repetition, whilst habits are strengthened by it; if the intensity of desire therefore were proportioned to the 'pleasure value' of its gratification, the desire for renewed gratification should diminish as this pleasure grows less, but if the present pain of restraint from action determines the intensity of desire, this should increase as the action becomes habitual." (James Ward in *Encyclopædia Britannica*, vol. xx., p. 79.)

But pleasure and so also desire often increases with repetition. One who tastes champagne for the first time may receive slight pleasure. The next time he dines out he will, with image of his previous experience, have slight desire for champagne. As experience is repeated his pleasure and desire may increase to ecstasy and passion. But habits not obviously pleasure-yielding, as the morning chore to the country lad, will be desired after intermittance; the country boy homesick in the city longs in the morning for the familiar scene and familiar task which was a source of aversion at home. We painfully miss the customary, even the painful customary, for thereby the conservative tendency of nature and organic activity is broken up. Desire arises for relief from this pain, and the habitual is so far regarded as pleasurable. Thus desire is in proportion to the "restraint" only so far as the restraint is painful, and thus relief appears pleasurable. Thus the desire for the habitual has, like other desire, its basis in prospective pleasure.

That the analysis of desire as regards representation of pleasure is still an open question certainly marks the psychology of feeling as very backward; that here is a 200most common and prominent psychosis, whose simplest analysis is not yet agreed upon, shows how far we yet are from a standard of subjective verification. I have expressed my own opinion that both the evolutionary standpoint and special analysis indicate a distinct emotion at prospective good which is best denominated by the term desire. This is a purely psychological result, and has absolutely no reference to ethics. "Pleasure" has such an inevitable ethical tinge that a purely scientific denotation would be useful. The "good" is a better, but also objectionable term. That then the organism should foresee and image the good and should have a feeling about it which should stimulate will to its appropriation and realization is a psychosis of utmost value, and one which is in all psychism above the lowest an extremely common phenomenon. This does not assert that desire in all its lower range is a seeking for pleasure, an extremely late conception and endeavour; but it means that as perception is of things in their experience values, so representation also, as giving the basis of desire; but a conscious hedonism is still afar off.

The general function which desire subserves in stimulating advantageous action is obvious. As anger and fear are primarily useful emotions in view of potential pain and harm, so desire in view of potential pleasure and benefit.

The function of desire in stimulating advantageous action is obvious. Desire answers to potential pleasure and benefit just as anger does to potential pain and harm. It is a correlative and supplement of fear, and in general the more one fears a thing the more one desires the opposite. When sailing I desire fair weather in proportion as I fear a squall. Desire is the very spring of life and progress, and when desire is extinguished the will to live ceases, and psychic life declines and dies. Fulness of desire is fulness of life, and the largest mental life is 201that in which desire, constant, multiplex, and far-reaching, is strong and dominant. Desire seems thus to be a permanent factor, and, though there is a pre-desire period, no post-desire age seems to be indicated in psychic history so far.

Somewhat as to the analysis of desire has already been intimated in touching upon its origin and function, but we are now to study its elements more in detail. The very young infant certainly experiences hunger pains in almost its initial consciousness; but it is only gradually that the need felt leads up to presentation and representation of the needed thing, and so to desire. Hunger with it, as with all organisms, sharpens the wits, and leads to knowing things, interpreting them, and acting definitely toward them. Through touch it first comes to appreciate object, and object as food, a representative–inductive act. The earliest meaning attached to object is edibility, and this, indeed, indiscriminately to all objects, as we see that infants mouth everything. Gradually from this, or by dint of a good deal of unpleasant experience, objects are divided into edible and non-edible, the primitive classification of things.

From the consideration of any such simple example as the desire for food we determine that the first element toward and in desire is a lack-pain generating felt want, and so—and such common use of words is significant—we want, *i.e.*, desire what we are in want of. A feeling of need or lack is fundamental. Now sense of lack is more than pain from restriction or intermission, for it implies a measure of in-ground integrated experience with objects, a constant connecting of object with purely subjective experience. For instance, hunger and feeling the need of food, the craving for food, are not the same, for it is evident that to feel lack of anything with such a central pain as hunger-pain means that this something has often been conjoined with the pain experience. 202Hunger is primarily an organic uneasiness and gnawing pain which does not include any sense of object as of a food or reference thereto. Our subjective and objective experience have been so completely integrated, and feeling of lack and that for a very definite thing has become so ingrained in mind with pains, we feel so spontaneously and immediately need of *thing* in connection with organic pains that it is very difficult for us to realize a state where this connection has not been formed or is forming. But it would seem that the first hunger pains of the infant are of this primitive quality, and that need is not felt in connection therewith. It is only after some crude cognitions of bodies have been generated in connection with the feeding act and as guides thereto that on occasion of hunger pains there can occur the sense of lack of food object, a painful feeling of unrealization, at first very dimly representative, and so a craving, an incipient emotion. Desire rests then upon capacity to feel the lack of accustomed satisfying thing in connection with some form of perception or representation of the thing. When a satisfying object is missing, it must be *missed* psychically before desire can awake. The reaction when a customarily conjoined experience does not occur is a peculiar feeling in mind, a disturbance, uneasiness, a unique sense of loss and lack which is the immediate stimulus of desire. Hunger at first leads blindly to activities tending to satisfy hunger, but the satisfying thing—food—therewith becomes gradually known, hence thereafter when hunger comes there is struggle both to know and to act thereby. This struggle has impulsation from feelings of lack.

Lack pains then prompt to cognitive activities to find the thing lacked and desired. The first knowledge is that some things satisfy, and an appropriative activity is excited. The lowest organisms under impulse of hunger pains reach out after things, feel for them, and as soon as they sense the edible, appropriate it. It is quite evident that they 203exercise cognition only as driven to it, and then it is effort even for the simplest knowing. But what the first psychic facts are is hard for us to interpret, because we have progressed so far beyond them. However, we may well believe that the general form of primitive consciousness is akin to what we have when dozing or half awake. The realization of things is dim indefinite, and it is only as pains of considerable severity are felt and as the psychism gains in capacity for pain that particular knowledges and particular needs and desires are accomplished. After having repeatedly sensed something—as a soft vegetable form—in connection with bodily pain as hunger and with the feeding activity as allaying hunger, a renewal of the pain from organic conditions will give, not merely purely subjective pains, but also, as the pre-associated cognition of thing and the allaying of hunger is not experienced, there arises as reaction a vague sense of lack which may lead to equally vague desire. A vague uneasiness and restlessness which knows object and misses object only in the most general way is the lowest basis. A study of some case of waking from a doze by reason of hunger would give the original formation of desire as involving lack sense. Here a purely subjective pain gradually intensifies till it wakens a very general objectifying, and we feel need of undefined something, which soon becomes specialized, when fully wakened, to need of something to eat, and finally as need of some particular usual food, as bread, meat, or milk, which is then desired.

Pain from restriction or intermission of some organic activity, as the digestive and assimilatory, may then lead to sense of lack and desire for object which is unrealized. However, craving-desire as implying sense of loss, of something pleasurable missed, is not organic, but is mere reflex of organization. It is not progressive, but conservative; it does not initiate, it merely keeps the organism to its accustomed level. This is the limited range of appetite. 204Craving rests on past evolution. However, we have to explain the origin of those activities which, when intermitted, produce such distressful results. We must first acquire the liking before we miss what we like, and tastes uniformly originate through effort, and all pleasurable activity is built up by painful volition as urged by direct pains or by desires. Desire then is more than craving. Craving as based on organic lack is satiable, desire is insatiable. We desire what we have never missed and modes of experience we have never attained. We, who have never had a gold watch, desire one, and having received one, we lose it, miss it, and so desire is reinforced. All the progressive activity of the human world originates in desire, as ambition, or as desire of truth, virtue, etc. Here we do not miss what we are accustomed to, but we are forming habits, which will be the basis for cravings with descendants. For instance, one who now does not miss beauty of art, but is ambitiously striving to appreciate art, may come finally—or at least his descendants—to miss art, and so to crave it. But for the time he has no art craving, only an art desire. Of course all desire in the craving form, or in the higher desire form, involves a missing actualization. All desire is extinguished in realization. But this obviously does not destroy the distinction of desire as based on craving, a spontaneous resultant from integration, an intermittence of habit, and desire as itself integrating habit-forming emotion.

However, with the lowest psychisms, we may perhaps suppose it unlikely that representation does ever become definite enough for desire, except when in direct sensing of a thing, as, for example, in a touch perception. The psychism is impelled to touch activity by its subjective pains and simple, undifferentiated lack pains. It does not desire a food through the representation of it brought up by hunger, for such representation of things in their potentiality is probably not originally stimulated directly 205by subjective feelings, though with man, for instance, we know that hunger and other simple feelings will provoke representations of foods, which foods will be desired; and particularly in famine the most lively representations of feasts occur, and thus there is a strengthening and defining of desire. Thus in famine there comes a greater and greater urgency to action as its necessity becomes greater. The vivid representations of foods become through desire—though there may be no sense connection with food—a mighty force for self-preservative action.

Yet primitively desire probably awoke only after some sensing was accomplished, not the mere subjective pain, but the touch perception awoke the representation, for it would seem the original *status* that representation occurs at first only with correlated presentation. Thus it is that the simplest psychisms are driven by their pains to achieve a touch or some sensing of a thing before they interpret it as food, and so desire it; that is, things must have a food meaning attached to them through actual sense appreciation of them as such, before they can be directly instanced in pure representation as foods. Hunger leads us immediately to think of food, but this ability to directly represent food is based upon having thoroughly learned certain things as food by repeated direct experiences. A savage who has never seen or known of bonbons is presented with a box of them, and he may receive them with indifference, but a bonbon is placed in his mouth, whereupon he says, "it tasted so good, I want another." Such is the genesis of desire when pleasure quality is attached to thing, is learned by experience. The visual and tactual experience is actively conjoined with pleasure experience, so that seeing another bonbon, he represents its pleasurability and so desires it.

Further, the relative presentations and feelings must be mentally correlative, the connection must be more than phenomenal series of several forms; there must be an 206active connecting psychic process as basis. You are told to open your mouth and shut your eyes, and a bonbon is dropped in; the taste will at once give rise to a revival visual presentation, and if a person holds up before your eyes a fine bonbon, saying, "look at this," there may occur revival taste experiences. But the immediate basis of desire is not here, for if psychic process stopped here, there would be no higher elements; these can only be accomplished by a definite bringing up and attribution of subjective quality to the thing. You represent its possible pleasurableness on the basis of past experience, by the action of the inductive instinct, a complex process. Here revival is not an active correlating, but is self-contained, lying isolated by itself, and unfruitful till its revival character is recognised, and it is actively wrought into experience. That is, integrating act is presupposed in all desire.

The way in which revival becomes the basis representation is hard to trace, but in many cases it seems to be connected with certain physiological activities. A revival form implies correlated physical functions, as when the sight of a peach causes the taste pre-experienced therewith to be revived, and the mouth waters, as if in actual deglutition. As the reacting and assimilating process is carried on without any real thing to be acted upon, there comes a physiological reaction, which in turn gives rise to peculiar psychic affections, and specially the uneasy feeling of lack. The unreality and mere revival character of the revival experience is ultimately recognised, and representation becomes possible, and idea of pleasure as both experienced and experienceable is evolved. Thus an unsubstantial revival, where the thing is sensed in one form only, but thereby re-awakening other associated experiences, as in the case of merely seeing a peach, leads finally to know the thing as a potency; I taste, but after all I taste nothing; hence I am led to perceive the thing as a sign, 207as unrealized in its pleasure significance, but realizable. How we attain sense of reality and unreality we discuss in chapter on Induction, but with special reference to desire we add here an illustration. When engaged in reading on a hot day, I have feeling of discomfort, and then spontaneously arises image of a wonted bathing place, I have the image of moving in the clear, cool water, but at once recognising the unreality of the image, I long for realization. I, when heated, have so often seen the water, and plunged in it, that the presentation of mode of relief has become firmly associated with the discomfort, so when it organically returns, presentation revives, and its unreality known, desire rises. One not accustomed to bathe, but to taking lemonade when heated, will have visions of lemonade and desire therefor. One who is just forming some habit of relief will not have spontaneous images, but must call them up. Desire also will be purely general, "Oh! to get rid of this heat." Specific desire, as founded upon a definite image of realization, is primarily the result of active association of definite object and mode with a given pleasure-pain state. The realizing the image as unreality, as suggesting an actualization to be wished for, is learned from rude experience with present sensations and perceptions quite at variance with the image. Thus, that the vision of water is unreality I know by seeing the room before me, touching the chair, sense of painful heat unrelieved, etc. An image of actualization barely of itself does not include desire. I may conceive that I can image myself moving in water without any emotion therewith connected, but as matter of fact, this never occurs; all our images of actualization carry some desire value. Even bare phantasy, as imagining myself living on the moon, is not without a tinge of desire or aversion, for the origin and growth of imaging has been so bound up with desire, and is for desire as life function that some desire tendency is retained even in the purest flights of imagination. 208It becomes increasingly evident that such a simple and understandable expression as, "I want that peach," implies a great complexity of psychic process which is hidden from us by the summarizing facility of language. Emotion is evidently far too complex for full analysis. Its complexity is such that we may well hesitate to attribute it, as is so often and easily done, to the lowest psychisms. Since desire includes a measure of self-consciousness, and also of consciousness of pleasure, it seems improbable at first sight that such elements should exist in certain low consciousnesses where primitive organisms seem impelled by desire. However, though this *a priori* view has weight, it must not be allowed to be of supreme value. Yet when we fairly interpret a very simple case, as when a dog scenting and seeing meat on a shelf, is said to desire it, and so to spring for it, we certainly imply a complexity of mental activity, which might by many be thought quite beyond the power of even a very intelligent dog. We have at least the following factors:—

1. Simple scent or vision of the thing; bare presentation or representation of object.

2. Either a definite bringing up, or a mechanical re-occurrence of past pleasurable associated feelings and sensations, or both.

3. Sense of unreality.

4 Feeling of lack.

5. Pain of lack.

6. Sense of pleasure potentiality of the thing, which implies—

(*a*) Idea of pleasure.

(*b*) Idea of personal experience thereof, *i.e.*, some egoistic sense.

(*c*) Sense of experience as in time past, as experienced.

(*d*) Sense of time as future as implied in sense of the experienceable.

2097. The longing, yearning, peculiar desire quality as feeling mode.

8. Desire pain.

In the first place then, the object of desire, the *desideratum,* is not the object as such. We do not desire things merely as such, but only as far as they are significant of experience. Presentation does not, at least normally and originally, ever end in itself, but it is always connected, and connects with pleasure-pain experiences. Desire begins by being vague as to its object; under slight pressures of pain we want something, but we know not what; we have dim, undefined longing, but the indefinite object is always a possibility of experience, a centre of pleasure-pain potency. At the first stirring of hunger pains, we have a vague uneasiness and sense of lack, with a most general idea of object and longing toward it, and suffer the pain from hunger. We may be physiologically hungry without feeling hungry, and so may have a desire of thing in general to remove pain before the pain is felt and recognised in its particularity as hunger pain. When hunger comes, or, primitively, is achieved, then we want something to eat; and as this feeling intensifies, the craving becomes more and more definite as to object; bread, etc., is wanted, and in famine hunger there is the most particular representation, as of certain dishes formerly eaten with great relish. Lumholtz, wandering famished on a Christmas in the wilds of Australia, thinks of the puddings in his native Norway. The evolutionary significance of this increasing definition of object in desire is obvious in that greater definiteness and accuracy of self-preservative action is thereby assured.

As far as the nature of the emotion desire goes, it seems quite indifferent whether there is presentation or representation of object. I desire equally, whether I actually see the bonbon on the table or when I merely represent it—see it in my mind's eye.

210Primarily then, and always, even in the latest evolution, as tendency at least, the desire is for the pleasure in the object, and desire is excited by every representation of the pleasurable. If one says, "I can look upon pleasure without desire," we may well question whether there is really personal pleasure represented. Dancing, card-playing, wine-drinking, may be pleasures which do not attract me because I do not care for them; and by such a statement we indicate the practical parallelism of pleasure and desire which is forced upon common introspection. If you care for it, it is a pleasure to you; if you do not care for it, it is not a pleasure to you; such is the result of common observation, and a very just conclusion so far as I can see. To excite desire, we naturally suggest the pleasurable. One person persuading another to go to a party says: "I know you would have a good time." When one answers, "I know that I would have a good time, but I dread the trouble of getting ready"; here is a conflict of desires in which desire of present ease and comfort may overcome desire of future pleasure. We may, indeed, assert that one cannot honestly say, "I know it would be a great pleasure to me, but I have no desire for it." When such a phrase is used, it can only mean that the pleasure is interpreted as belonging to the generic class of pleasures, yet not a pleasure to the individual in his present conception, or else its contingency, implied by "would," is so great that desire is practically *nil*.

And if the pleasurable is always the desirable, the desirable also may be said to be only the pleasurable. The martyr in his most eager desire for a painful death, fixes his mind, not upon the pain as pain, but upon the enduring it successfully, and the triumphant pleasure, also the satisfaction of the reward of martyrdom, and the pleasure of suffering for right and the approval of conscience; these and many other factors influence him.

Desire is *at* pleasure, not *in* pleasure, and thus contains 211pain, especially as implied in the preparative factors, sense of unreality and sense of lack. A bonbon may be so cunningly imitated, that placed in the mouth it feels like a bonbon, yet not tasting so, the painful sense of unreality and loss occurs. There is a painful waking up to the fact of non-realization, much the same in quality as that which we suppose to have happened in the original genesis of desire. The pleasant hallucination is broken in upon by actuality not fulfilling the psychic co-ordination pre-established under more favourable circumstances; and this occurs in early psychisms on a wider variety of occasions than in later development. That I am not tasting the bonbon I see on the table, this fact *per se* does not pain me. I take it as a matter of course in an order of nature already well learned and completely acquiesced in. But with infantile and lower stages of evolution generally, the lack of immediate correlation seems highly painful. Seeing has directly developed in immediate connection with a tasting, and the seeing without tasting seems by its very nature as disquieting as the feeling in the mouth the artificial bonbon without being able to taste is for later experience. It is through the negations of customary coincident impressions that anticipation and desire become forced by the exigencies of life. The early psychism is limited in its adjustments to a very few simple coincidences, but in the struggle of life in complex nature there comes disruption of these primitive co-ordinations, and sequences become apprehended, and meaning is discerned in things. This disruption primitively occurred most easily when there was direct opposition to the usual course of sensations. Just as when mouthing the imitation bonbon, we apprehend most quickly and easily non-realization when it tastes sour rather than sweet. By realities continually breaking in upon the common course of psychic association, the significance of things is gradually apprehended, and to see a thing is understood 212not merely as coincident with other sensations and perceptions, touching, tasting, and pleasure-feeling, but the thing is cognized as centre of pleasure potency, and so can become object of desire. Experience loses its self-contained simplicity, and is forced in the struggle of experience in a complex environment into some definite understanding of things, and into a feeling for them or at them, and not merely a feeling from them. And so a world of desirables and aversibles is formed.

If no pain was felt in the experience of unreality and lack, if there was mere passivity, desire would not be generated. This pain of loss spurs the mind to achieve desire, and desire enables the organism to attain the advantageous. At length a conventionalized world of desirables so formed, and certain significances, become so inground into experience that they seem often to be instinctively and immediately recognised by the individual, anterior to any personal learning by experience, as in cases of instinctive fear of, and desire for, certain objects.

While desire is attained at the incitement of pain, it is in itself a painful mental act. The emotional going out toward the *desideratum* is in itself a painful mode of consciousness. The feeling I have for the bonbon which I see and desire is, so far as desire, painful, yet negatively and comparatively, it may be pleasurable in that this psychosis may supplant one more painful still. It may be said that desire is painful, and also lack of desire, or *ennui*. But mere desirelessness is not *ennui*. *Ennui* is a feeling of lack and loss, and so a feeling of desire, but a peculiar kind of desire. It is desire for activity, when by a morbid *status* there is no desire moving to activity. Lack of desire and interest in things may be painfully revealed to some active natures, but to the great majority of psychisms it is a pleasure state. As far as we can judge, the undesire of the cow leisurely chewing her cud in a warm corner of the barn yard is supreme felicity. A 213state of desirelessness, complete yet blissful, occasionally visits even the consciousness of the nineteenth century busy-body. But the normality of desire for human adult consciousness in general is apparent to all. One who loses all interest or desire loses hold on life. Thus desire is life, and even when it is sought to extinguish it either as dictated by a philosophical maxim or by religious and moral scruples, on account of the innate selfishness of desire—Madame Guyon, for instance—yet desire is sure to intrude, and must as a desire to destroy desire. So whether we *would* fly, or *would* reach desire, we thereby desire. We may uproot or cultivate certain kinds of desire which thereby become objects of aversion or desire, but the effort to extinguish desire as general fact of psychic life involves either a psychological indefinite *regressus* which is never desireless, or else it means the extinction of consciousness itself in any grade above the lowest.

A further element which appears in all desire is some measure of self-consciousness. The representation of the experienceable implies some representation of experiences. In constituting the world as the sum total of the experienceable, we imply an ego-consciousness, and that objectifying as psychic act is correlated with subjectifying. Desire, like all other emotion, implies a subjective reference. We see clearly that the psychic act expressed by "this is the food," and as such the precursor of and ingredient of desire, means an identification with past personal experience. A similar act is performed, no doubt, by animals very commonly, though not expressible in speech, yet in measure expressible, as in the cluckings of a hen to attract the brood to some seeds. In various ways the *desideratum* is suggested to the mind, and in view of it, both in the identifying as having experiences, and the longing to experience, some consciousness of personality is implied. This in early forms of 214psychosis is, no doubt, meagre and indefinite enough, but not more so than its correlate sense of object. When a strange object is presented, as when a famished traveller finds a new house, identifying effort is instinctive; he at once seeks to understand it, and gropes through his past experience to determine what has been its life significance for him or other persons, and so what will be. What is thus done in the full light of reflective consciousness by man, is done in a summary and imperfect manner, generally by psychisms, as preparative to making the object a *desideratum* or *anti-desideratum*. The assimilating and integrating, the knowing, never exists without some appreciation of subject, because integration is not only of something—objectifying act—but also to something—subjectifying act. Things are from the first apprehended only in their immediate egoistic significance, and also very early as centres of possible sensations which become a matter of fear and hope, desire and aversion.

Desire is certainly a very extensive psychic *genus* including many varieties which are noted by common introspection, and which are even denoted by special words. A wish is a momentary act of desire, longing is intense form; ambition and aspiration are desires for higher order of objects, as contrasted with desire for food or dress. The kinds as distinguished by object are numberless since any object may be desirable, and the realm of the desirable is coincident with the realm of the knowable. In the course of evolution we become aware of things and of states of consciousness, so that by feeling about them—having emotions—there may result the advantageous action with reference to them. This order, not consciously apprehended of course, is the natural order of psychic events, and one which, in tendency at least, always appears, even in latest evolutions. The desire to know what is the full experience value of things, curiosity, 215is an early acquirement, since complete cognition of object is obviously of the greatest advantage, especially to the weaker animals, as deer, who act wholly on the defensive. The very strong can afford to be largely indifferent to their environment.

With reference to intensity, we can place forms from a positive to a negative pole. Thus with a famished man desire for food is first intense craving, becoming with continued eating moderate desire, then feeling of satisfaction, then of repletion, then negative, as aversion in passive form or satiety, then becoming active as disgust, and intense as loathing. Content, or desire satisfied, is not desire extinguished, rather it is an equilibrium wherein desire and its function are in continual equalizing action. When desire granted means all desire extinguished, with beings of any high tendency to activity *ennui* is the result. Here, as Schopenhauer notes, wish for a wish develops. Even in complete pleasurable quiescence, there is desire for its continuance, which is only saying that there can be no complete quiescence short of coma, or else of a state where reality has never broken in, and experience is wholly unformed where the being cannot anticipate or note change. Pure and absolute content never occurs, and as a matter of fact never will, the point of transition in the desire gamut, in passing from positive to negative, being like a mathematical form, unreal and theoretical. When positive desire ends, negative desire springs up immediately, just as in the pleasure-pain gamut, where the indifference point of transition has no existence in reality.

Desire in any of its forms may take on an altruistic, disinterested phase, though much that is taken for altruistic is only apparently or partially so, being really due to self-extension. If you take an interest in anything, it becomes interesting to *you*, it is a matter of personal concern, and becomes identified with the self. Thus in 216*our* family, *our* town, *our* nation, *our* race, desire plants itself; it is in this personal extension of view that most of the pity, sympathy, and benevolence is exercised. A well-wishing and consequent exertion for humanity in general is very late, and still later is desire for animals as sentient beings having a worth of being in themselves.

The remarks we have made concerning desire proper, apply equally to aversion. We must bear constantly in mind that desire proper and aversion are really in psychic analysis, merely phases positive and negative of a certain definite mode of psychosis, hence we often use desire in this large and generic sense, which instances will be apparent from the context. Desire, like other emotions, is polar, and desire generic has its antipodal feeling in some form of active desirelessness.

As desire is naturally and originally connected with all perception of object, we find it closely allied with other emotions. While we must suppose that early desire is upon idea of pleasure, upon the idea of its realization to be attained, without any estimate of likelihood or unlikelihood of realization, which factor is slow in evolution, yet when through experience, sense of certainty or uncertainty is attained as to the experienceable, this psychosis—belief—has a marked effect upon desire, and is closely associated with it. Bare sense of the experienceable was sufficient to generate desire, but when the measure of probability of the experienceable actually happening is measured, we have belief, expectation, hope, and kindred psychoses, bound up with desire. The expression, "I hope it will be a good day to-morrow," indicates a wish that it would, *plus* some confidence that it will be a good day; "I wish it would be a good day, but I fear it will not," shows some lack of confidence in the realization of the event. Hope then equals wish, *plus* the intellectual element expectation, a desire for a realization *plus* some belief in it as actually to happen. A large 217share of learning by experience consists in the reaction of this expectation on the wish, in learning not to set our hearts on what we believe to be unrealizable or extremely improbable to happen. Wish also acts on belief, as is plainly expressed in the common phrase, "the wish is father to the thought." If belief tends to restrict or magnify desire, desire also tends to determine belief. Hope, as very commonly used, as when we say, "I hope it will turn out so," is a passive emotion, and does not appeal to the individual as self-determining the event. As the primary end of emotion is to incite the organism to determine its own experience, hope as passive seems a rather late evolution, as having only an indirect and general value by maintaining general pleasurable tone. The one who hopes it will be a good day to-morrow is in a better and more advantageous frame of mind than he who fears it will be a bad day, in so far as the events are equally beyond self-determination, and it is of no direct use to either hope or fear.

As to the range of desire we must then disagree with Aristotle and later psychologists, who suppose that desire is limited by the belief in the possibility of realization. Desire existed before this belief was generated; and while, after its generation, it may often affect desire, yet often it does not. I may wish for the moon as readily as the child to whom the notion of possibility or impossibility of realization is beyond experience. The unrepresentable only cannot be wished, and desire is bounded only by the power of conception and perception. Hope is a species of desire which has to do with belief in the possibility of the event or act: it is a joyful emotion connected with belief of realization of the pleasurable. This distinction between hope and desire in general is implied in the phrases, "I wish he *would* do it," and "I hope he *will*." The hope includes the desire, but the desire may exist without the hope, as we say, "I wish he would, but 218know he won't." Desire may be hopeless, but hope cannot be desireless.

Desire is vitally connected with ideation and volition, but properly it is the intermediate emotional moment between these, and not idea of pleasure—as James Mill—nor yet to be placed under will—Bain, James. It is neither phase of ideation or volition. Desire is neither idea of, nor striving after realization; it is not the idea of goal nor the effort to reach goal. I may have idea of a goal without desire to reach it—at least, analysis discriminates thus as separate mental stages—and I may desire to reach it without trying to reach it,—impotent desire, sometimes called wish. The striving is the consequent, and the idea the antecedent of the desire which is the emotion wave we emphasize by the word, longing. Desire is neither phase of volition nor ideation. Volition is properly effort at realization, and is stimulated by the emotion toward the realization ideally apprehended.

The relation of desire to will has been a fertile subject of discussion from Aristotle down, but we have to take up but a single aspect, namely, whether will and desire may with reference to the same object be contrary or distinct. Take the example of contrariety mentioned by Stewart. I wish a certain man not to do a certain act, but yet I persuade him to do it at the request of a friend. If I say I will persuade him, though I wish him not to be persuaded, this merely implies that the wish to oblige my friend overcomes the aversion to persuading the man. And, in general, apparent cases of conflict of will and desire may be resolved into conflict of desires. Hence the phrase, "I will do it, though I do not want to do it," is inaccurate or rather an incomplete analysis. We should always add, "because I have some extraneous and stronger desire." A box of bonbons is hung in a room at a height to be had by whomsoever will jump and reach it. In any party of persons there may be some to whom the wish for 219ease, the disinclination to jump, overcomes the inclination for the bonbons, so that this volition does not occur, others who jump even against this disinclination, the desire for the bonbons being the stronger desire, and others, very active, who jump without feeling any disinclination to the act. Conflict of desires is a common and almost constant state with many minds, and the evolution of man has been mainly through conflict of desire in sacrificing an immediate to a future good. In lower minds with so little self-consciousness and consciousness of a consciousness that they do not grasp conduct as a whole, there is a simple alternation of volitions flowing from the desires of rival goods, till one by its intrinsic force dominates with some permanence. These are the creatures of impulse, unreflecting and unself-directing by principle and reason. Higher minds realize *their* situation and consciously bring in higher desire or motive; they form rules and principles of conduct: they become ethical beings, having self-control and self-direction.

Desire is based by Mr. Bain on hindrance and opposition to activity, on "a bar in the way of activity." This is true if we understand it to refer to sense of unreality and of lack as connected with an apprehension of thing where the thing is really absent from the usual correlation, and hence physiological activities are checked. We have in the previous pages discussed this, but this is not Mr. Bain's point of view. The three elements he emphasizes are: deficiency, idea of pleasure, and the hindrance. Thus, he contrasts the prisoner who looks out on a bright day and longs to take a walk, with a perfectly free man who looks out on a fine day and freely follows his inclination to walk. However, it appears to me that both have desire, and that in the same sense both are moved by the motive, though only one is free to attain the action. So if I get thirsty in a waterless desert or in my room with a jug of water on the table, the bodily sensations will equally lead 220to desire. The conflict in desire is between state actual and state conceived, and not between will and restraint. Mr. Bain remarks, "If all motive impulses could be at once followed up, desire would have no place." (*Emotions and Will*, p. 423.) But desire is itself an original impulse, and is more or less an ingredient in all emotion impulse; and it is plain that emotion impulses as implying representation are the only ones which can be "followed up." Where every wish is gratified as soon as formed, as with a petted child of rich parents, desire still remains in all its characteristic quality. Such an one, however, by having only the momentary pleasure of completed realization, misses the joys of realizing, and loses all that happiness which has been defined as sense of progress. If every wish were gratified as soon as formed, if every representation of pleasure was immediately followed by realization, desire would still exist in all its peculiar force. The moment of gratification is always second to the moment of desire, and a Fortunatus with his wishing-cap cannot possess in absolute coincidence with the wish.

It may be objected that Tantalus' desire is certainly a form where hindrance is the main stimulant. When one is continually hindered just on the point of realization, desire is intensified, but this intensifying is very largely due to the increased definiteness of presentation or representation, and to the increase of confidence in the event. To tantalize is to bring before one an object of strong desire into the clearest prominence and seemingly certain attainment, yet to constantly withhold it.

We have spoken of desire as an impulse, and we would include all emotion as impulse, for to impel is its function and action. Impulse is the will side of emotion as interest is its intellectual side. If I fear a man, this is my interest in him and impulse from him. True, we speak of being driven by "blind impulse"; but emotion cannot be blind, it can only be kindled by object imaged. Anything 221which actuates the will may very broadly but wrongly be called impulse, for impulse strictly connotes an emotion wave undirected and issuing at once in action. Where unforeseen ends are served, as when a hen driven by sensation of heat sits on eggs, we commonly but wrongly denominate it impulse. Without some representation there is no emotion and no impulse. So when standing over a precipice I say I have the impulse to throw myself down, this means that the depth wakens in me image of falling and an awful desire to realize the image, which impels the act. If I am merely giddy I will fall, but if I have the emotion-impulse I will throw myself down; I am not impelled by dizziness or any sensation, but the term denotes emotion as desire or fear.

For the ordinary human mind desire seems in general a spontaneous and instinctive act. We do not make an effort in desiring, though desire like other mental functions undoubtedly arose in struggle. Originally this psychosis was a stress and strain activity; it was a rarely achieved emotion, just as the emotion of pleasurable appreciation of Beethoven's music or Michael Angelo's sculpture is for most minds a rare uplift of psychic force. Knowing as compound of presentation and representation and as involving emotion and volition, is, with us, within certain limits, an habitual spontaneous act of mind. I feel the pain from cold, without sensation of cold, as bare pain, as undifferentiated feeling of discomfort, I then feel cold, I feel cold object, I desire warmth, I will to draw near the stove; here is a progressive series of correlated psychoses which are constantly occurring in a spontaneous way in ordinary experience. But this psychic structure which operates so easily is really the outgrowth of ages of psychic evolution wherein the separate steps have been achieved and the correlation established only by the severest *nisus*.

Associations are first achieved in experience established 222by numberless reiterations before there is spontaneous tendency to re-occurrence, this is the law of psychic evolution to-day, and is the only clue we have to the past. The evolution of mind is not and never has been a mechanical process, but its basis is in pure feeling as stimulating volition. Paradoxical as the expression sounds, yet in a sense it is true that the organism has *learned* to know and to feel thereupon. It may even be that in the course of psychic ages with certain species of animals some emotions may become innate, and such advantageous psychoses as fear or desire may occur without any integration through individual experience. The new-born chick, when it hears the note of a hawk, is said to show signs of fear, though what actual psychosis occurs, if any, seems almost beyond our power to know. The whole process may be reflex nervous action, a mere closed neural circuit being affected. It is no doubt true that all long-continued, often recurring psychoses tend to so embody themselves in a neural combination that the given activities are carried on in a sub-conscious and finally in an unconscious way. It is very probable that much that we take for emotion with lower animals is reflex or semi-reflex action; yet it is likewise true that there is, as a matter of advantage in struggle for existence, an inherited instinctive tendency to certain emotions, to certain kinds of fear and desire, and there may be a distinct awareness of the potency in things, which has never been individually realized. In its every transaction with things the young organism may act by reflex action or by inherited emotional tendency. How far either or both enter into the first individual experiences is a matter for the psychology of the future.

The general function of desire in life is obvious; it is the most potent factor in conserving and extending life. Far back in a paleozoic psychic period life was below desire; but once originating under the pressure of the 223struggle for existence it has since developed into the most manifold and complex forms. Human life is the outcome of desire, and the human being is *par excellence* the desiring psychism. As the moving factor of humanity history is its record, and present human organization, faculty, and achievement is its product. Desire, as the force to realize, to convert seen potency into actuality, the idea into reality, is now in the very highest examples of psychic development an ever increasing power, and no prospect of a psychic stage to be reached beyond desire is intimated in the present course of normal development. The tendency toward extinction of desire, when it does occur, appears always as pathologic or retrogressive symptom. It may be the dream of a philosopher or of a cult, but with Schopenhauer himself desire was a most forceful factor, and the devotee of desirelessness by very reason of being a *devotee* to an object, desires it, namely, the state of desirelessness. We may desire to extinguish certain desires, and succeed in accomplishing this, but to desire not to desire, as general act, is a psychological contradiction in terms. A very low vegetative psychic status without any desire is possible, but all teleologic activity implies desire, hence extinction of desire can never be attained as an end.

Desire moves the world and is the core of psychic being. Deprived of definite desire, we long for it, and if every wish were immediately realized, we should desire delay in gratification. The amount and value of life is measured by the quantity, quality, and effectiveness of desire. Orton characterizes the Indians of the Amazon as "without curiosity or emotion," which must, however, be taken only as relatively true, but yet marking them as extremely low in the psychic scale.

Education then, is a process of stimulating desire, of leading to ambitions and aspirations. As what is imposed on consciousness without desire is a hurtful burden, the 224true pedagogic method is always to awaken the wish for knowledge and power before it is granted. Desire as interest is assimilating power, and without it there is no mental growth. The art of education is the art of stimulating intellectual, æsthetic, moral and religious desires, and of providing for their progressive gratification with the best arranged and most suggestive material.

CHAPTER XIV

SOME REMARKS ON ATTENTION

The term attention is, like feeling, a word of extremely doubtful and variable import. Like feeling, attention may be used as denominating any stage of consciousness, or it may be restricted to some more or less specific form. As affections of the organism all psychoses are termed feelings; viewed as subjective-objective acts, a content being attained, consciousness as such is termed attention. We are said to be attending when we have any activity of mind, when we have anything in the mind or before the mind. When consciousness has something in it, consciousness is attending, whence attention means consciousness acting. But what is consciousness inactive? Nothing. Hence consciousness attending, used for consciousness acting, is a pleonasm. Consciousness, by virtue of always being conscious of something, does not need the word attention to qualify it.

The attention of consciousness is called, attracted, or engaged, when any mental act occurs, whether a pain, pleasure, perception, or whatever form it may be. When the mind is occupied with anything, *i.e.*, is active, it is thereby attending to the thing. If I am conscious, I am, of course, conscious of something, hence attending to that something. But all these expressions are incompatible with a purely psychological point of view. In psychics, as opposed to physics, the thing exists only as perceived and in perceiving, *esse* is *percipi*; the object or content of 226consciousness exists neither beyond consciousness nor in it; it is consciousness and consciousness is it, it is nothing more than objectifying fact. Consciousness does not, like a pail, have contents, but it is merely a name for the sum of activities we term conscious. Such a phrase, then, as, attending to something, may be radically misleading. We do not have both consciousness and a field of consciousness, a presentation field. A tolerably constant part of human consciousness is an activity which is a constituting a world of external and internal objects. This objectifying activity, which may or may not be object for higher activity—apperception or attention in one sense—does not, however, persist and subsist as a more or less mechanical *continuum*, as Mr. James Ward and that school maintain. Still the word attention may in a vague and general way denote both the realizing force and will effort therein of every act of consciousness. But yet as thus a general term for certain aspects or general qualities, it is liable to misconstruction, and we do not propose to employ it either as denoting any act of consciousness as such, or any aspect thereof.

Attention may also denote dominancy in consciousness. When any one factor is pre-eminent, we say the mind is therewith attentive. When any element has a marked ascendency, so that all others are much feebler and subservient, thereby is constituted a state of attention; as when sight perception monopolizes consciousness in an eagle watching for prey, or hearing commands all the mental powers of a deer listening to a strange sound. However, practically all states are in reality complexes in which some one factor is and must be dominant, and this universal phenomenon of dominancy scarcely deserves the specific name, attention. Consciousness is always more or less concentrated in some single channel; the factors in any state of consciousness are never perfectly equal in intensity, and so are never in perfect balance. But attention 227is not this fact of dominancy, but rather that of consciously sustained dominancy, as we shall note later.

If attention is not a proper term to denote simple dominancy, may it not denote that complete form, engrossment, or absorption, where one element predominates to the exclusion of all others, and so occupies all of consciousness—that is, more exactly, is all the consciousness—and also tests the capacity for consciousness to the full? The fixed idea is an instance in point, and in a certain way also preoccupation or absent-mindedness. Still, in this last there are manifold elements and often great complexity—*e.g.*, train of thought—hence dominances of different forms, but yet a persistence of a certain mode with consciousness running at its full capacity, and the result being that the general trend is not easily altered. In cases of fixed idea and brown study we say, "his attention is fully occupied," which means nothing more than his mind or consciousness capacity is fully taken. I do not see that we gain anything by using attention in the same sense as these two general terms, mind and consciousness, which are surely sufficient. Further, when one "loses himself in a subject," the power of self-activity, and hence power of real attention, is lost. Mental activity which has slipped beyond the control of will is not in any true and high sense an attending, nor is attention good term for consciousness at saturation point.

Again, attention is often used to denote consciousness in its change aspect. When a new consciousness comes in and supplants a former state, we say, in popular but misleading phrase, "it takes or attracts his attention," as if attention were entity rather than activity. But when we say that change of consciousness is change of attention, we really add nothing; it is an identical proposition. Attention does not qualify consciousness, but is merely synonym for it.

Still again, may attention designate intensity, or some 228certain degree of intensity? We may say of one, "he was looking inattentively," or of a fixed, intense gaze, "he was looking very attentively." A strong vision is thus opposed to weak as an attention. As all psychoses have some degree of intensity, they are thereby acts of attention, if we reckon from a zero point, or a more or less large number of consciousnesses if we reckon from some fixed degree of intensity. But to call a psychosis, because of its intensity, or because it has reached a certain degree thereof, an attention, seems an unnecessary procedure. Nothing is gained by describing an intense psychosis as an attention, and certainly intense pains and pleasures hardly come under the term. Nor yet are intense cognitions, merely by reason of the intensity, properly states of attention. Fixed ideas are commonly intense, yet there is no true attention, as we have before intimated. Cognitions which come as intense must be marked off from those which are intense by reason of a self-determined self-consciousness intensifying. The essence of attention is intensifying act self-regulated. To be sure, intense presentations are given as such only by an heredity *momentum*, from past ancestral intensifyings; their *impetus* is on the basis of past cognitive exertions. Presentation intensity, and, indeed, all mental intensity, is originally and fundamentally volitional; the act had its force solely in will power; but in late phases psychoses which originally required intense exertion rise spontaneously and have a strength and persistence apart from volition, and so the word attention does not rightly apply to them. Thus also we can solve the problem that Mr. Ward states when he says, "How the intensity that presentations have apart from volition is related to that which they have by means of it— how the objective component is related to the subjective—is a hard problem; still there is no gain in a spurious simplicity that ignores the difference" (*Mind*, xii. p. 65). But "objective component" and "subjective" 229do not enter into the question; cognition does not arise as a given, as forced and determined from without, but it is rather at bottom a mode of volition. Still attention is not then cognition intensity in general.

If attention is not any form or quality of mental activity in general or of cognition in particular, we must find its essence in volition—as, indeed, has been intimated in the immediately preceding pages. Attention is properly the will side of cognition; it is cognitive effort. Considering attentively, looking attentively, listening attentively, mean cognitive efforts in thinking, seeing and hearing. Here is a cognitive experience which does not simply happen, but is definitely brought about and held to. There is intensifying act by which the given cognition is held and kept in dominancy. The word attention must, as a psychological term, be extended to denote, not merely modes of cognitive effort prominent in man, but all cognitive exertion of whatever grade. It will include all will-tension in all the senses—olfactory, gustatory, muscular, etc.—as well as visual and auditory.[D] A dog scenting game may be as truly attentive as a waiter listening to your order. So far as the smelling by the dog is merely instinctive, that is, heredity survival, there is no real attention; the mental activities are not efforts of will-attentions—so far as they occur spontaneously and inevitably. But when, as we often see, a dog is somewhat baffled in scenting, it plainly puts forth cognitive effort, it exerts its cognitive powers to the utmost, there is that strain and stretch which the word attention literally and naturally suggests. As soon, in fact, as the labour point is reached in any mode of cognition, here is attention. All toil and work is attention, as a definite exertion of will including some cognitive element. The labour of life is attention, is minding or attending to business. Attention is thus will effort in maintaining and intensifying a mode of cognition.

D. See also my remarks in *Psychological Review*, ii. p. 53.

230Concentration of attention is then, we may now remark, a redundancy, as we make attention equal to concentration. To say his attention was concentrated upon a certain subject, is equivalent to saying his mind was concentrated. Sometimes, indeed, concentrated attention may mean intense attention or concentration, but some concentration being always involved in attention, it is a confusing and inaccurate phrase.

In a more restricted sense, attention is not merely any will tension in cognition, but only so far as self-consciousness is involved in all the exertion. We must sharply distinguish between this attention as willed activity and as simple act of will. Willed cognitive activity denotes cognition determined upon and consciously accomplished. The willing in the knowing act may not be will to know. Willed cognitive activity, when not against the will, when including choice and acquiescence, is in the true sense voluntary attention—attention voluntarily, freely, willingly performed. The term voluntary is not the proper correlative of spontaneous, but rather volitional, while non-voluntary must be set over against voluntary. In self-conscious attention of any kind there must be consciousness of the tension, and consciously exercised effort in delineating and maintaining cognition. In this narrow sense attention is conscious furtherance or hindrance of cognition. Effort is consciously put forth in some particular cognitive form; there is a self-limitation by the mind in cognitive process. In short, attention here equals cognition consciously constrained.

As to the relation of attention to subject, we remark that psychology as the science of mental phenomena, rather than science of the soul, is not called upon to imply a subject as in any wise attending. Yet we use, and use inevitably, substantive forms and personal pronouns, but while it is impossible for science to desubstantialize language, yet it must be ever on its guard against the 231delusions of language. It is a common impulse to explain activities by referring them to agents, to describe attention and all mental acts as being what they are by reason of the actor, the self, or *ego*; but science in this, as in so many things, inverts the common order; the agent is made by and of activities, and not the reverse. Agent or subject is no more than a congeries of manifold interdependent activities. There is, and can be, no fixing of the mind by the mind: the word, mind, being used in the same sense in both cases. When I say, "I fix the mind upon something," this means for analytical psychology, that in the complex of consciousnesses which are unified by an *ego*-sense, there occurs a will effort accomplishing a perception. This purely dynamic interpretation is the method of all science which cannot accept inexplicable essences and agents as explaining anything. Attention is not to be explained by an attender, but it is a mode of activity in that collection of activities which we term organic life with conscious process. So even attention, as self-conscious exertion, is not to be interpreted as an agent which is conscious of itself in exerting; but we consider it as volitional activity with consciousness of self as manifold complex of objects vitally connected with will effort. Self-consciousness does not necessarily mean a self conscious of itself.

It is obvious from our discussion thus far that we do not accept the common division of attention into spontaneous and voluntary, which means for us no more than spontaneous and voluntary—more properly volitional—cognition. So-called spontaneous attention is the displacing of one consciousness element by another without any will effort; there is no displacing or placing as will activity, but cognitions appear, persist and disappear by an inherent force. When in deep study the noise of a whistle may spontaneously "attract my attention," as the phrase is, but this denotes no more than forcible change of 232state. There is nought in the new act but the sensing the noise of whistle; there is no real attending activity, no will effort at either promotion or inhibition. However, we must grant that most cognition contains a volition element. Absolute zero or negative value as to volition is but a momentary and comparatively rare phenomenon in normal consciousness, where self-possession and self-direction in some measure is almost constant. In the case of noise of steam-whistle suddenly breaking in upon a student, there is quickly attention—either positively, as listening to quality, or to detect direction of sound; or negatively—true *in*attention—as inhibiting and disturbing element. When one is made "wild," or distracted, by noise, then his mind is occupied unwillingly, indeed, yet there being no real promotion or inhibition, we must term the state *unattention*. Another form is where we give up in despair, and passively suffer the annoying noise. In both cases we neither stimulate nor repress, and so both are emotional unattentions. On account of the pain-pleasure nature of all experience, there is even here, however, some will attitude and tendency, some favouring or retarding act, though it be wholly impotent in effect.

Just when a cognition rises to attention point, just when volition with effort becomes prominent factor, this is a difficult and delicate problem. However, according to the relative prominence or obscurity of volition element, we must divide cognitions into attentions and impressions. In the variety of human cognitive activity there is a constant flow of cognitions which are one moment being strengthened to attentions, and another, weakened to impressions. With volatile persons cognitive life is a kaleidoscopic congeries of rapidly experienced impressions and attentions. Will darts in and out with marvellous velocity, now vivifying some, now others, in the stream of cognitive activities determined by pleasure and pain interest. With all of us there is a manifold complex *continuum* 233of cognition, a general non-attention knowing of external world and *ego*, which we continually carry with us. Into this field of exertionless cognitive life will-effort penetrates now to one point, now to another, seizing upon and enlarging the most interesting and significant facts. As I am sitting in my chair, I am dimly aware, without will tension, of a large field of varied objects, any one of which I may emphasize, attend to, when incited by sufficient interest. Practically exertionless awareness is a constant *substratum* for developed consciousness; here, in the world of habit, it is always at home, and moves with great ease and smallest friction; but the process of learning, the work of adding to mental possessions and enlarging the *totum objectivum* and *totum subjectivum*, this is attention for complex consciousness.

We must note this, that attention is any general alertness toward cognizing, though no actual cognition be attained. Cognitive straining without result is truly a form of attention. A man listening for a sound is equally attentive with a man listening to a sound. It is not necessary for an attention to have something to attend to. Attention is effort at cognizing as well as in cognizing. The stupid boy is often the most attentive, the most strenuous in cognitive effort, yet there may be little apprehension. In fact, we must recognise that in cognitive, as in muscular activity, effort may be excessive, and defeat its own end. When suddenly awaking in the night we often strain sense to the utmost, but with no result; nothing is heard or seen. In this, as in some other cases, we must notice that attention is not necessarily delineation. While generally a particularizing effort of cognition, attention may sometimes occur as mere general cognition stress.

If attention consists in cognitive effort, whether successful or not, what is the nature of the effort to attend? A student says, I try to attend, but I cannot; I cannot hold 234my mind down to anything. Professor James remarks, "In fact, it is only to the *effort to attend*, not to the mere *attending*, that we are seriously tempted to ascribe spontaneous power" (*Psychology*, p. 451). But it is obvious in such phrases attention means simply cognition, and may be substituted for it, whereas we have just pointed out that attention is both the effort toward and in cognizing act. Literally interpreted, then, the problem is whether we can make an effort to make an effort at cognizing. In great lassitude or exhaustion we lose control of ourselves, we are unable to exercise volition either as attention or otherwise. We recognise and lament the fact to ourselves, we feel our powerlessness, but I hardly think we do ever really make an effort at effort. At the very first stage of recovery from such state of utter non-volition, the will act is always toward definite sense adjustments, or in holding to and promoting certain thoughts and representations, and we thus have real attention. The utter rout of psychoses, which once possessed us, we now conquer and control for our ends and interests.

Attention to attention is obviously distinctively different from this phase. We can and do attend to attention as psychic fact. An act of attention cannot, indeed, attend to itself, but the volition act in consciousness of consciousness, as consciousness of some attention act, is very properly an attention to attention. If I am looking attentively at a man, I cannot, by the very nature of attention, be simultaneously volitionally introspective of, *i.e.*, attentive to the looking attentively. When actively sensing light, I cannot at the same moment attend to this attention, because attention is always concentrative of will. To be volitionally conscious of light is one moment, and to be volitionally conscious of this light consciousness is another moment. The attention attended to is not in process at the same moment as the attention. This does not deny that we have simultaneous spontaneous introspection 235of attentions. Introspection, like sensation, perception, ideation, is attention only so far as it is effortful.

In his recent treatise on psychology Professor James discusses in an interesting and suggestive way the relation of ideation to attention, maintaining that "ideational preparation ... is concerned in all attentive acts." Attention is "anticipatory imagination" or "preperception" which prepares the mind for what it is to experience. Thus the schoolboy, listening for the clock to strike twelve, anticipates in imagination and is prepared to hear perfectly the very first sound of the striking.

It is undoubtedly true that in the form of attention we term expectant, where we are awaiting *some given impression*, there is a representing, antedating experience, which may be a preparatory preperception. But with a wrong imaging of what is to be experienced there is hindrance, as when in a dark, quiet room we are led to expect sensation of light but actually receive sensation of sound. Very often, indeed, our anticipations make us unprepared for experience. Further, the experiments adduced by Professor James from Wundt and Helmholtz are in the single form of expectant attention, and we must remark that in these experiments the reagent is also experimenter, and this introduces a new attention, consciousness of consciousness, and that of a peculiar kind, which complicates an already complex consciousness. In general we may say that experimentally incited consciousness is artificial, at least as far as it feels itself as such, and for certain points like simple attention this tends to vitiate results. Self-experimentation or experiment on those conscious of it as such may mislead in certain cases, and must, so far as this element of consciousness of experiment is not allowed for. In physical science things always act naturally, whether with observation or experiment, but in psychology observation, other things being equal, is more trustworthy than experiment.

236In all cases of expectant or experimentally expectant attention, the attention does not, however, lie in the expectancy or in the imaging as such, but it is merely the will effort concerned in these operations. Yet as we may expect without effort, and preconceive without volition, attention is necessarily involved in neither. A perception or a preperception is an attention only as accomplished by will with effort, but only an unattention when purely involuntary. Professor James's use of attention as preperception brings us back to the common idea of attention, as any consciousness which cognizes something. This is so inbred in thought and language that it is most difficult to avoid using the term in this sense. Many psychologists, like Mr. James and Mr. Sully, frequently mention attention as a will phenomenon, but they do not treat it under will, and they constantly return to the cognition meaning. Höffding, however, treats attention under psychology of will. Attention as the exercise of will in building up and maintaining cognitive activity, is naturally treated under cognition; but it is on the whole safer and better to discuss attention under will so as to keep it sharply distinguished from the presentation form which it vitalizes. I have endeavoured to hold the term strictly to this sense, yet it is not unlikely I may sometimes unwittingly countenance the common confusion, but trust the instances will be few.

When we have, then, a case of expectant attention, we must distinguish the attention in the imaging from the attention in the actual cognizing. It is, indeed, true for us almost invariably that cognitive strain without immediate realization is incentive to ideating. In listening in the night in vain for a sound we hear in imagination many sounds, and we form preparatory ideas of what we are to hear. Sense-adjustments call up a train of sensations in ideal form. But it is obvious that low intelligences which have no power of expectancy or ideation do yet 237really attend. The very first cognitions and all early cognitions by their very newness and difficulty were attentions long before ideation was evolved. With low organisms, as cognitive power extends only to the present in time and space, immediacy of reaction is imperatively demanded, and every tension of cognitive apparatus is immediately directive of motor apparatus, so that suitable motion is at once accomplished. The cognition, though dim and evanescent factor, is yet powerfully energized, and so a true attention. Always with lowest sentiencies, and often with higher, pain is suddenly realized without anticipation, followed quickly by attention as strong effort to cognize the nature and quality of the pain-giver and so to effectually get rid of pain-giver and pain.

Preliminary idea, then, cannot occur in early attentions and in late attentions, it is by no means necessary. It is said that we see only what we look for, but it must be answered that seeing commonly happens without any looking for. The kindergarten child, Professor James to the contrary notwithstanding, is not confined in his seeing to merely those things which he has been told to see and whose names have been given him. A child continually asks, What is that? and is quick to discern the new and strange. He accomplishes a wide variety of attentions without ideas and gives himself almost entirely to immediate presentations.

To be sure, every one sees only what he is prepared to see, only what is made possible for him by his mental constitution as determined by his own pre-experience and the experience of his ancestors, but this does not signify ideation. Every cognizing is conditioned by the past, but this does not call for a reawakening and projecting in ideal form at every instance of cognitive effort before any real cognition is reached.

In fact many, if not the most of our attentions, are merely intensifyings of some present cognition, of some 238cognitive psychosis which has simply come or happened. Take the instance of attention to marginal and retinal images; this certainly does not always imply pre-perception, the forming of an idea of what we are to see, though in the cases mentioned by Professor James it may. For example, I was writing the above seated with my profile to the window when I became suddenly aware, through the physiological agency of a marginal image, of a moving object to my right. This perception of bare, undefined object was spontaneous, a pure given; I exercised no will in attaining it, and so the state of cognition was not an attention. However, by attending, by intensifying the cognition by will effort, I perceive that the indefinite object is a man walking on the sidewalk, who is of a certain height, clothed in a certain way, etc. I do not trace the least ideation in the whole process; the slight attending as act of will did not imply any anterior or posterior idea or representation. The reason for the will act was the intrinsic interest of movement, and this intrinsic interest arises in the fact that moving objects have had for all life a special pleasure-pain significance; the moving object is the most dangerous, and so motion perceived has become ingrained in mind as a special stimulant of attention. This habit of attentiveness to things in motion survives and continues for cases where it is of no use and even of harm; thus, in the present instance, it diverts me from my work. It is obvious that attention often occurs in the same way for other senses without preliminary idea.

Is there such a state as negative attention or active inattention? Is will activity in cognition always positive merely, and never existing as direct repression or weakening of acts? To some psychologists negative attention means only that certain elements in a consciousness are overshadowed by the dominancy of some single factor; that, owing to the limited capacity of mind, many elements can exist only in enfeebled form beside their stronger 239neighbours. If the life blood of mind, will, is largely absorbed by some particular form or mode, all other forms must suffer in consequence.

It is, of course, obvious that the amount of will force which is put into some given cognition is potentially or actually withdrawn from other factors which then, however, are more justly termed unattentions than inattentions. But is the withdrawal of energy attained only by transference? May it not be attained by direct repression and suppression? When we wish to weaken some particular cognition, is it to be done only by specially energizing some other cognition? It would seem on general principles rather strange that we can, under stimulus of interest, increase our energizing of any given cognition but cannot reduce it except indirectly by transference. This would mean that the sum total of actual will force remains constant as far as subject to voluntary control, and it is only by subdivision into many channels that any actual diversion is secured. Will force may be withdrawn and transferred, but not an atom of it can be directly suppressed. But can I not directly repress a troublesome thought or a painful sight? If by a great effort of will I keep my eyes closed to some horrible but fascinating sight, this is a true active inattention, the exactly opposite exertion to holding my eyes open and fixed upon my book for reading when very sleepy, which process is always termed attention. When our energy is going in some comparatively undesirable way we often do simply switch on to another track, but often also we shut off steam and reverse. Instead of direct promotion or indirect inhibition there is direct inhibition or often both forms of inhibition combined. We may, under pressure of interest, directly weaken any cognition, untensify, check and reduce the will effort involved by immediate relaxation. In putting ourselves to sleep we relax with effort, we reduce and stop all attentions. In awaking we often go through a reverse process. The 240attitude of any cognition is either by and through will, or with comparative indifference and no intervention of will or with will directly against it, which three states we term attention, unattention, inattention.

Negative attention is then, I think, a real activity, a will force which directly hinders and crushes out the unwelcome in consciousness, while positive attention is will force vitalizing and strengthening the pleasant. In conflict of interests these forms are complementary, and attention is here a double will-effort, both the effort at withdrawing energy from one point, and the effort at applying it in a new point. In most cases attention is both resistance and insistance. Even in simple forms the natural tendency to inertia constitutes a constant counter interest to any particular activity-interest. Attention then is always resistance to this natural inertia *plus* the direct energy in effecting the particular activity. But in advanced consciousness there is always a multitude of difficulties in the way of specializing cognition, a great variety of distractions to be resisted, all which, added to the definite exertion required in the special work, makes the ordinary attention in human consciousness a very complex affair. A student engaged on a mathematical problem is incessantly driving out distracting thoughts and positively fixing his mind upon the problem. Resistance is manifold, according to the speciality of the task—the more special, the more distractions—and the direct concentration is also a real and direct activity.

We may then, I think, see the importance of both positive and negative acts in attention. As counter to the theory that positive attention is the only real form, we might plausibly argue the opposite, that it is only the reverse side of negative attention. If we shut out all but one element from consciousness, do we not thereby bring that one into bolder relief and so indirectly strengthen it? May not all intensification of cognition be thus but an 241indirect result of negative attention? No, for even when all distractions are kept away, there is the inherent difficulty of the act *plus* the inertia, the general disinclination to effort. Positive attention may rarely appear as practically pure, and rarely also negative attention. Consciousness may sometimes consist of merely pure will tension as keeping off all defined activities; and persons of great will power sometimes achieve this in putting themselves to sleep. Consciousness is a blank field, tensely kept, but perfectly so only for a very brief time.

As to the origin of attention, it must arise with cognition itself. The past act of cognition was, as we have seen (p. 61), a powerful will act, an achievement through struggle, and therefore an attention. The history of cognition and of its ultimate development into the highest forms is a story of incessant and fierce competition in the struggle of life. Man's power of sense, perception and thought is an inheritance from an immense deal of will effort by untold millions of ancestors. The necessities of existence compelled an alertness, a general cognitive strain, which effected progress and discovery, the attainment and integration of new and most valuable forms of experience which have been handed down to later generations. The earliest cognitive life is then almost entirely attentive; cognition does not *come*, it must be *attained*. Gradually, however, some low form like general sensation is so integrated, and requires less and less attention, till it *comes*, is *given*, with comparatively no effort, and a state of unattention thus appears in consciousness. The child repeats quickly, easily, without attention, the evolution of the past, and this spontaneous re-enactment continues up to the full point of hereditary integration. Without effort the child is carried on at the incitement of instinctive inherent interest up to a certain comparatively high grade of experience. But heredity *momentum* gradually ceases, and if there is to be individual progress, attention must come in. Thus, 242intellectual education is fundamentally a developing of attention. Conscious control of cognition, both positively and negatively, becomes more and more efficient, and the progress of the race is dependent on exceptional attention in exceptional individuals—geniuses. Attention becomes more and more limited and specialized, and a minute subdivision of labour results.

Now, primitive attention is not as Mr. Ward, for example, would make it, a primordial fact of mind, but as a cognitive form of will or will form of cognition—it is essentially secondary. However, Mr. Ward, in his article in the *Encyclopædia Britannica*, makes a peculiarly advanced form of attention the initial fact of consciousness, namely, by the non-voluntary act of mind being conscious of changes in itself. But mind is not at first a something which is inevitably cognizant of its own experience, but it merely is a state, does not have states, and is not consciously aware of them as such. There is, for instance, pain, but no consciousness of the pain as fact of experience. Mind is not primitively a something acted on, reacting, and cognizant of these self-movements, but merely effortful will activity attaining snatches of cognition at the pressure of pain and pleasure. It seems, indeed, tolerably plain that apperception is not necessary to consciousness as such, and the general law of evolution from simple to complex leads us to suppose that consciousness was not at first with any apperceptive process. Changes, whether as occurring or as being brought about, did not imply an apperception taking cognizance of them. But however this may be, certain it is that apperception, as consciousness of self-change or as consciousness of consciousness, must as a form of cognition arise in will effort like any other forms, must be a real attention, not a so-called non-voluntary attention. We do not see any reason why this form of cognition should be an exception to the general law that every step of consciousness 243is an acquirement and achievement determined by the struggle for existence.

The relation of attention to feeling has already been touched upon, especially as related to interest. Attention, like other volitions, is aroused by feeling, primarily as direct pleasures and pains, secondarily by the ideal forms of these, that is, interest. Low organisms are incited to attentions as simple sensation-cognitions only by present or immediately impending pain or pleasure. Direct pain does not interest or include interest in itself. There must be, not merely pain, but cognition of it as element in experience, before there is interest, which is always *in* something. Interest implies representation, the sense of the value for experience of any given thing. What pleases or pains interests only so far as perceived as pleasurable-painful; the thing perceived as source of feeling, or as in any wise related to it, arouses interest. "I am pleased or pained," does not equal, "I am interested"; but only so far as I have cognizance of the object, pleasing or paining, am I interested in it. The interesting is what touches my interests, what affects my experience, what potentially reaches or touches me. It is obviously to the great advantage of the organism that pleasure-pain object merely perceived should move, excite, or interest, which brings in attention to the thing, and so fuller knowledge and preparedness for action. Interest, then, is practically equivalent to emotion. "It interests me," is equal to, "It arouses my emotion." The interesting picture, book, man, animal, etc., is that which awakens emotion, and thus incites attention. What affects me or moves me, interests me. Interest is generally used to denote favourable emotion of rather low intensity, as when I say, "He interests me"; but as a psychological term it may well be used in the broad sense to denote any emotion so far as it stimulates attention. The function of interest lies wholly in its effect upon attention, it is always a feeling stimulant to the will act of cognition. I do not 244exert my cognitive powers unless I have some interest at stake.

There are, of course, many degrees of interest. Often interest is so slight as not to rouse attention, being too weak to overcome natural inertia to will effort or unable to deflect will as bent by some conflicting interest. A lesson is to be learned, but the interest, often extrinsic, does not rise to attention point till possibly a few minutes before recitation. The interest, fear of failure, may then be sufficiently strong to induce very vigorous attention, and within a certain range the stronger the interest, the stronger the attention. Yet at a certain point of intensity emotion begins to derange will activity and to hinder and even destroy attention. Fear which has become fright extinguishes attention. Self-controlling power of attention is lost in a flood of emotion. Yet ungovernably intense emotion is no longer properly termed interest, which always implies cognitive power. Interest is properly comparatively mild emotion state, which includes definite cognitive element. But interest may be not only at or below attention point, but it may be of such an intensity and kind as to do away with need of attention, securing a spontaneous, or practically spontaneous, cognition. Thus, my interest in a book may at first be insufficient, *i.e.*, practically *nil*, to constrain attention in any degree; it may become so strong that I make constant cognitive effort, and finally, as it becomes profound and absorbing, I cognize without any attention. When anything becomes sufficiently interesting, interest acts of itself directly upon cognition, which is then performed without attention. Interest frequently increases to the spontaneous cognition point, carries cognition in it; but we must remember, nevertheless, that all cognition had its origin in attention. Interest acquired and become habitual demands less and less force of attention, so that our customary interests finally awake cognition without any attention act. If given cognitions always required the 245original will effort,—attention,— intellect could not progress, delicate and far-reaching reactions could not be initiated, for they could have no basis. The force of inherent hereditary interests makes itself felt throughout all advanced psychic life. A survey of the cognitions of any single day would show us that by far the greater number are by this type and degree of interest. The common cognitions and adjustments of every-day life in walking, sitting down, and in matters of routine, are mostly of this type.

It is tolerably plain that the relation of feeling to cognition cannot be expressed by any single formula, and it is certainly far from true that sensation or other cognition is inversely as the intensity of feeling. If feeling, either as simple pleasure-pain or as interest, is the incentive of attention, which is the primary measure of cognition; then intensity of cognition is directly as intensity of feeling for a certain range, and this is also true where attention has lapsed. The law of inverse ratio applies only when feeling has risen beyond the point of highest efficiency, when there is over pressure, and mind runs wild beyond self-control and attention. Then we should, of course, find at a certain point, if we could make exact measurement, geometrical decrease in cognition for arithmetical increase in feeling, but ratio would constantly change. The centre and spring of any high psychic life is interest, and as interest increases intellection and volition increases *pari passu*. In cases of decline, where interest or capacity for emotion is lost, psychic life as a whole dissolves and disappears. On the contrary, the progress of mind is in the strengthening and extension of interest.

Interest leads to attention in the forms mentioned, but it seems also a mode of attention when, at the bidding of interest, we not only promote or inhibit some cognition, but some particular feeling. In a fit of anger we may be prompted by prudence or conscience to forcibly and directly restrain and abate it. I may similarly maintain 246an amiable frame of mind as opposed to crossness. To repel a fit of anger of course implies repelling the representations which enter into the angry emotion, and so it is that the repressing or stimulating all emotions, by reason of their representative nature, necessitates a will effort with reference to the cognitive element, and thus an attention.

It is commonly believed that attention to a feeling intensifies it—that the more we attend to our feelings the stronger they are, and the less attention we pay to them the weaker they are. A soldier wounded on the field of battle heeds not the pain in the excitement of the conflict. But the truth is in this case that he has no pain so long as he feels none, and that he does not attend to the pain signifies simply that pain does not become a psychic fact, but is wholly physiological, and so not a subject for psychological discussion. This is a case of the confusing use of attention for consciousness in general which we have before criticised. Very often, indeed, such an expression as, "The more he attends to his pain the more he has," means simply, the more pain he has the more he feels, an identical proposition. But we must also discriminate between attention in a feeling and attention to a feeling. I work myself up into a passion by strenuously dwelling on representations involved in anger—this is an attention in a feeling; but attention to anger would be self-observational effort. The former does not involve consciousness of the feeling, the latter is nothing more than strenuous consciousness of the feeling. Men are often angry without being conscious of it or but dimly so, and attention to the feeling would consist in intensifying by will effort this consciousness. When a person says, "I was mad and I knew it," he asserts the distinctness of the acts and that the first does not always imply the second. This cognition originally, like all cognition, required volition, and it is still subject to volitional control and emphasis, that is attention, even in advanced consciousness. Attention to a feeling is cognitive 247effort in attaining or strengthening consciousness of feeling, hence is but a mode of apperceptive or introspective effort.

We must distinguish sharply then between the observing act and the observed feeling, between a cognition of consciousness of pain and a pain consciousness, and we must note that attention may be either, neither, or both. Apperception has become such a habit with higher human consciousness that it is commonly exercised without attention, and so has seemed to some as a necessary fact of all consciousness, an anthropomorphism, which seems to us erroneous. When we are conscious we are generally conscious that we are conscious; when a man has toothache there is not only pain, ache, toothache, but consciousness of this as fact of experience; but this does not establish apperception as fact of all consciousness.

Is it true now that the more we are conscious of a consciousness the less we have of the latter? Certainly the more conscious we are of it does not imply having the more of it, though we may say with truth that within a limited range the greater and intenser the consciousness, the greater the facility for consciousness of consciousness. A mental fact must have a certain definiteness and prominence before it is clearly and easily cognizable. However, speaking of the effect of apperception upon the consciousness apperceived, it must be evident that it is always a minifying and not a magnifying. Consciousness is self-divided when there is both experience and consciousness of experience, hence a loss of force for the consciousness cognized. A feeling self-consciously felt is weakened thereby. The feelings we are most conscious of are of comparatively low intensities. In very intense feelings we lose or forget ourselves: we do not know what we are doing or feeling.

If now we make the consciousness of consciousness effortful, it is plain that we diminish the consciousness 248cognized in still greater measure. A consciousness of consciousness cannot be forwarded except at expense of general mental capacity, and so as diverting force from the act observed, whatever this be. Attention to a feeling must then on general principles diminish the feeling, and that in a marked measure. The psychologist who is always twigging his own consciousness to find out what is going on there must often be surprised to find nothing there. It is astonishing how fast feeling disappears when we begin to examine and analyse it. The emotion fades the moment we turn attention to it. We find that in psychological matters as elsewhere that we cannot have our cake and eat it too. We murder to dissect. Apperceptive effort is never intensification in the consciousness cognized, but cognition and pleasure-pain feeling as a consciousness cognized lose in force, just as in the body, an undue exaltation of one function is always a depressing of others by withdrawal of force. The more conscious I am of my fear the less I fear. While this law of withdrawal of force is obviously the case when consciousness is at its fullest capacity, yet it may be said that apperception in other phases acts as stimulant to waken latent forces, just as in the body stimulus of one function is often stimulus of all, though we doubt that apperception is original and permanent function in consciousness. But still in such cases it is a new consciousness which is stimulated and strengthened and not the consciousness which is being cognized, and still more then is there decrease in the latter. A given feeling is never increased by attentive consciousness of it. When a feeling is said to be intensified by attention to it, we may suspect either inaccurate analysis or misuse of terms. This, of course, does not deny that within a certain range *immanent* attention increases pleasure, etc., for example, the more actively we taste an orange the more taste pleasure we get.

We note in passing the very interesting psychological 249paradox that the more we view ourselves the less we have to view, the principle of which has been set forth above. We know well that the very reflective and self-conscious have little personal force and individual quality. Moreover the self-conscious stage in youth is precisely the period when there is the least real self to be conscious of. A strong multiplex mind is rarely very self-observant.

Finally we have to remark upon the way in which attention may be divisive of cognition. Boswell makes Dr. Johnson to say, "If we read without inclination, half the mind is employed in fixing the attention; so that there is but one half to be employed on what we read." But admitting the necessity of intrinsic interest, this does not do away with attention. Attention hinders rather than helps cognition only when it becomes wearing strain, as in reading when much fatigued. But attention as fulness of vigorous normal will activity gives a force and value to cognition which it would not otherwise have, and often makes its very existence possible. The greatest, most significant cognitions in the mental life of any individual are those which are achieved at the top of endeavour. Real knowledge as advancement and acquirement is always the fruit of long training and attention.

The act of attention is painful and therefore is not exercised by lower organisms, at least, only under absolute necessity. Often the pain from attention is so great that the individual prefers to suffer than to exert himself cognitively and so help to remove pain-giver. It is only under the greatest pressure that new knowledge and new ideas are acquired, and the history of mind shows a series of *tours de force* achieved only in moments of direst need. The strengthening and the holding of cognitive powers to a given point by effort of will is peculiarly distasteful and painful activity. All minds tend toward inaction or toward the regions of effortless action where overwhelming interest carries them freely along. Attention, while the 250most advantageous of actions, is yet most irksome and painful. It would seem to us at first blush that if pleasure and not pain had attached to the attentive act from the beginning, the evolution of mind would have been accomplished in the merest fraction of the time actually required. It would have been the difference between going down a steep incline rather than up. Why progress should only be realized through painful effort and struggle is a problem which has vexed the thought of man throughout history but upon which psychology has little light to throw. Our present concern is to simply emphasize the fact that cognitive act as attention is always painful, and if the act of cognition is performed without pain we may promptly deny this to be an attention. This is, of course, far from asserting that all cognizings with pain are attentions.

CHAPTER XV

SELF-FEELING

Popular and scientific observation agree that a very interesting and important phenomenon in consciousness is the sense of self as involving such feelings as pride, shame, self-satisfaction, and self-disgust. And the evolutionary psychologist is bound to consider self-consciousness in its rise and development as a life factor. What is its significance for life? How and when did it arise as answering a demand in the struggle for existence? Further, the psychologist is bound to clearly define and analyse the self-sense as psychic fact, to understand just what it is, as well as what it seems. The nature of the self-sense must be carefully studied by introspection, and its elements and quality determined. However, the psychist has nothing, of course, to do with the self which is sensed, an inquiry which belongs alone to the metaphysician.

Self-consciousness has been throughout all our discussion assumed and implied as factor in emotion life. Object is not merely perceived, for this in itself has no life value, but is at once interpreted in experience terms, is self-related, and emotion arises and stimulates suitable will-response in bodily activities. Thus all response to environment through cognition of environment means with sense of the environment as its own. Thus, and thus only, is sense of environment rendered efficacious, for bare objectivity, which signifies nothing, has no value for life. 252Under the conditions of existence in the struggle of life object cognition could not originate because it has no function. The theory of natural selection then requires that object and subject cognition be regarded as complementary psychic factors, coincident in their origin, and developing in strict correlation.

This corollary from the theory of natural selection, implying a self-relating act in all cognition under the condition of struggle for existence, is seen to be a likely hypothesis so far as we can judge from the action of low psychisms. Any one who closely observes animals must recognise that self-interest determines their cognitive activities and in turn is roused by it. The alert listening and looking of a squirrel is obviously impelled by fear and awakens fear. The object perceived is constantly interpreted for its experience value, that is, there is constant self-reference. This is the type of all cognition under natural selection, *i.e.*, where use dominates.

Assuming then psychism as mode of adaptive reaction, we see the necessity for the correlation of the sense of self with the sense of things. An experiencer blind to self, who has no awareness of self, but merely blindly strives, has little advantage, for it possesses no self-directivity and no power of intelligent action. Its adaptation is purely general; to be specific adaptation it must appreciate differences in environment in their differential action upon itself, an appreciation of the objective in subjective terms. It is probable then that the first knowledge was the apprehension of thing as painer and then of the thing as pleasurer. A discrimination of the two is attained, probably tactile, as hard and soft. The subjective import of the thing is at once realized from these signs.

It is obvious that the origin of self-consciousness must be placed very early in psychic life. With organisms which have but a few flashes of consciousness during their whole individual existence, whose whole experience is a 253mere sum of separate pleasure-pain thrills and blind efforts, there is neither sense of objectivity nor subjectivity. These very lowest psychisms have experience, but no sense of experience; pleasures and pains possess them, but they do not possess these. But if mentality arises and progresses solely by virtue of its function in saving and profiting the individual living organism, if the end of psychosis is this self-conservation of the bodily whole in its vitality, there is an imperative demand for self-cognizance in order to self-care. Under the law of struggle and survival of the fittest, the organism which does not look out *for itself* must go to the wall or be in the lowest grade. Self-conservation is closely linked with self-sense. Hence the individual very early acquires some sense of itself in its environment, and so acts and conducts itself. Thus under adverse forces it learns to know itself, to realize its own place and power, and to feel fear, anger, and so to appropriately respond to any environment. Thus is secured manifold and special response to multiform conditions, whereas in the organism which has only pure subjectivity of pain the response would be uniform.

The condition of an *ego* being sensed or known is, of course, that there is an ego to be sensed. All experience is an individual's experience, is personal, but this does not constitute egoism as an experience. The experiencer must have experience before he can know himself as experience centre, that is, there must be experience before there can be experience of experience. But the amount of consciousness and integration thereof which is required for self-cognizance is probably very small. The dynamic organic whole of psychic life, which we denominate *ego*, has almost from the start self-consciousness, and grows by self-integration. By the conjoint interaction of subject and object cognition with feeling and will elements egohood or personality is gradually developed to the largeness which we see in the human mind. Experience which 254does not self-integrate is scarce worthy the name, and it is noticeable that we usually associate self-consciousness with the term. "Having an experience" signifies a self-related psychic fact. Given the first germ and experience constantly returns upon itself and self-develops. It anticipates itself, experiences the experienceable, and so serves life. A psychic individual without sense of his own individuality is practically undiscoverable and impossible. It is perhaps not too much to say that psychically egohood really begins when experience cognizes and organizes itself; the self is made by the sense of self. At first only an occasional achievement upon a very meagre basis of psychosis, the self-sense rose only through intense pain and effort, but has now become so built into experience that, with human minds at least, it seems constant and spontaneous factor. Just what this means we have to note when we come to analyze the self-sense.

While the ego-sense is to be regarded as a reflection of experience upon itself, this reflection is far from being abstract, or general, or spontaneous. The self-sense is wrought out in the direct commerce with objects demanded by the exigencies of existence, a particular and concrete apprehension is produced. That is, mind is no purely internal development nor yet a mechanical impression. Development is forced upon it in a world of competition and danger, but yet this development is always active response. The self-sense then by which the individual becomes aware of its own activities and feelings as its own, originates, like all other new modes, by stress and strain as a most valuable psychosis in the struggle of existence.

The primitive self-consciousness is evidently naïve, that is, there is no consciousness of the self-consciousness. The low psychism is conscious of itself, knows what is to its own advantage, and is absorbingly selfish, but it is wholly unconscious of its self regard; so also with very young 255children we see an egoism which is perfectly unconscious and naïve, often humorously so to the observant adult who perceives the utter simplicity of its selfishness. The embarrassing self-consciousness of the boy and girl in their teens, a conscious self-consciousness, is not yet achieved. The immediate consciousness of self cannot by itself embarrass, it must be complicated with reflection and with cognizance of other *ego*'s; but later forms we do not need to discuss here.

In the simplest form of self-consciousness what are the necessary elements? and what is the essential nature of self-consciousness as psychic fact?

In the first place, then, what is the nature of self-consciousness as cognition? If cognition be awareness of object, what is self or subject cognition? Is subject merely a kind of object? Is self-consciousness a peculiar conscious mode, or is it merely of the same type as the general cognition of object? Of course we wish to consider such questions here simply in the light of psychic fact.

It is often considered that self-cognitions are really in no way unique, that the subject sensed is merely the individual's body or his mental powers. And it is undoubtedly true that subject is always some object, the subject cognition is apprehension of some object either corporeal or mental; yet self-cognition is never merely an object seen as object. The psychic act of self-cognition is a peculiar qualifying of the object cognition; the individual who merely knows body or mind has not self-sense, he must be aware of body and mind as his own. The essence of self-sense is not in the object as so perceived, but in the subjectifying reference. While the *ego* then is always constituted as object, *ego* sense as psychic fact is more than mere object cognition. The psychic self as object, as some mode or modes of consciousness, has naturally been emphasized. Thus the self may be defined 256as that which is subject to will. Yet the least reflection shows us that for self-sense this must imply *my* will, and so assume what it would explain. A consciousness of will act as effective psychic fact is not *ego* sense. A cognition of effort or *nisus* is not the sense of self save so far as the effort is known or felt as *mine*. And so in any other objectivist definition of self as psychic object, the self in its real nature as psychic act vanishes. Thus the consciousness of pleasure-pain capacity, while closely related to self-sense, does not make it, for we have to add that the capacity must be known as one's own. In every endeavour then to define or analyze the self as psychic fact we must either eliminate it or presuppose it, and this must be taken as very significant. It means at least that this *stating* it—being merely objectifying act—destroys the subjectifying which is its essence. The radical distinction and polar opposition of subjectifying and objectifying is therein suggested, and the difficulty of all fruitful discussion and scientific investigation, which is objectifying, is made apparent.

The objective cognition of a self can only mean cognition of an object capable of experience. Objects are thus discriminated into two classes—experiencers and non-experiencers, subject-objects and bare objects; but this is not self-sense whereby the experiencer directly knows *his own* experience as such, but merely sense of a self as any individual object experiencing. This objective definition of a self is simple enough. It merely asserts that any object which at any moment of its persistence or existence has a consciousness or experience of any kind is thereby a self. But this is obviously not a definition of the self and self-sense as psychic act, nor does it explain it. The scientific statement that individual objects exist as experiencers, and so are personalities, or *ego*'s, does not clear up the self-sense whereby the individual is aware of his own individuality as such. Egohood as selfishness in 257this objective sense, and ego-hood as self-experience, as a feeling and knowing myself, are quite distinct. To the question, What makes an object— this particular object, body with limbs and various organs capable of feeling pain-pleasure—what makes this *myself?* the only answer is relation not, be it noted, to experience, but to *my* experience felt as such. And what makes an experience mine is that I consciously experience it; not merely that I experience— that experience occurs to me, or in me, as objective fact—but that I consciously experience, subjectively realize the experience as *mine*; not merely as realizing experience as experience, but as *mine own*. This ceaseless circle into which we fall in trying to define *ego* is hinted at in various common expressions. A child even will often remark, "I did not do it, my hand did it"; "you did not touch *me*, you touched my *foot*," etc. That is, even the most cursory observation asserts that object in itself is not subject, that the me is not mine.

While, then, we must regard self-cognition as a *genus* by itself and as unanalyzable simple psychic fact, arising early upon a very slight basis of experience, and continually developing as most important psychosis for life, we may yet distinguish what is involved with it, what modes of consciousness it presupposes, and from which it yet is distinct.

We might speak of ego-sense as an experience knowing itself. But since cognition implies always a knowing and the known, an experience cannot, and does not, know itself. The consciousness knowing is never the consciousness known; and to speak of a consciousness as aware of itself is misleading and inaccurate. To speak of the cognizance of a pain as pain self-cognizant is an erroneous expression, for the pain does not know itself; but it is known by a cognition which is not it. To be aware of pain as such is awareness of consciousness, but is, interpreted strictly, in no wise self-consciousness. I may even 258speak of a self-conscious self-consciousness. This does not really mean what it directly implies, but can only mean a self-consciousness *plus* a consciousness of it as one's own; that is, the self-consciousness is not actually conscious of itself. Even if a consciousness could both be and know its being as an absolute, simple act, yet this would not be self-sense, an individual realizing its own individuality, but merely a single psychic act existing, and at the same time conscious of its existence. Self-consciousness is more and other than any consciousness which is self-conscious, if that were possible.

Consciousness of consciousness is not, then, self-consciousness. It is, indeed, conceivable that an *ego*, in objective sense, might know his own consciousness not as *his own*—the act of self-consciousness—but merely as consciousness, and he would thus exist as an individual, yet without subjective individuality. Yet, as matter of fact, consciousness of consciousness always carries self-consciousness with it. If I become conscious of a consciousness which is my own, I know it, not merely as a consciousness, but as my own consciousness; if I am conscious of anger, I am conscious of being angry.

Hume, in his chapter on Personal Identity, observes, "For my part, when I enter most intimately into what I call *myself*, I always stumble on some particular perception or other of heat or cold, light or shade, love or hatred, pain or pleasure. I never can catch *myself* at any time without a perception, and never can observe anything but the perception." This is a good illustration of a futile and mistaken attempt to absorb self-consciousness in consciousness of consciousness. Of course Hume was not the hypothetical *ego* which we have instanced as purely objective observer of his own consciousness; when he was conscious of any consciousness, as a heat or light sensation, a pleasure or a pain, he was assuredly, like other mortals, conscious of it as his own. The sense of mine-ness 259as psychic fact he should not have ignored, whatever might be his conclusions as to the *myself* But metaphysical psychology is always apt to swerve from fact.

The close connection of self-consciousness with consciousness of consciousness leads often to their confusion. Thus under the head "Illusions of Self-consciousness," J. M. Baldwin, in his treatise on the Senses and Intellect, says, "Of these subjective illusions we may mention *emotional illusions*, wrong estimates of our emotional states, as when an angry man declares that he was never more cool in his life." This instance is plainly an illusion of introspection, not of self-inspection; there is a mistake in the consciousness of consciousness. Wundt, in defining self-sense as perception of the unity of experience, falls into the same confusion.

It points to the fundamental value and place of these cognition factors, that when we say any one is conscious we imply them all. Thus I say of some one rendered unconscious by an accident, "He slowly recovered consciousness," by which I mean, became aware of himself and his surroundings with awareness of his own mental activities. He is consciously conscious, objectively conscious, and self-conscious. All this makes up for us being conscious, and is for cognitive mind such a simple organic basal movement as circulatory-nervous-motor function is for body.

An organism must, of course, have had some psychosis before it can become conscious of it, and of it as its own, and this primitive psychosis we regard as pure pleasure-pain series. But in the struggle for existence the organism is driven out of this subjectivity to cognize its environment as related to itself, to apprehend and comprehend and so to feel about itself—emotion—and so led to intelligent will activity as real self-activity. At the very first the organism has pleasures and pains, without knowing them as determined in itself by objects, but this primitive 260pre-cognitive stage is short, and most psychisms are certainly beyond it; they sense and notice things, bodily and beyond the body, as of experience value in pleasure and pain terms. At some most critical moment cognition first arose as triple movement, object—subject—consciousness knowledge. Just what may have been its original form it is most difficult to determine, but we may suppose it to have been a very weak activity, possibly expressible, as, "it hurts," object being simply pain centre. "It hurts," means object self-related, with consciousness of the consciousness, and this is our language expression for what seems to be an extremely common psychosis among many organisms. As simple pains were probably the first conscious phenomena, consciousness of pain was probably the first consciousness of consciousness, involving also subject and object consciousness. Not only to have a pain, but to be conscious of it as definitely objectively determined is decidedly useful attainment, which is finally inground in experience, so that it occurs spontaneously in highest psychisms. But it is only with a few of the highest human psychisms that consciousness object and subject are apprehended as general facts. Even by philosophers and scientists, subject, subjectivity, and object are not easily apprehended in their distinctness as purely general modes; it requires will strain to properly know them.

We have throughout sought the origin and place of modes of consciousness in function, and from this point of view we must view object-knowledge, subject-knowledge, and consciousness-knowledge as early coincident and correlative. Cognition springs up as a threefold mode, for in no single factor by itself has it life value. Pain, we say, forced the organism to work out to object as painer, cognition arising at once as triple activity. However, this does not imply that there is a constant knowing with, an apperception, that every consciousness is accompanied with a 261consciousness of it. Pains, pleasures, perceptions, etc., constantly engross the consciousness field without our apprehending them. Simple, common folk and children are rarely apperceptive, but yet they are eminently self-conscious, and consciousness conscious in all their life of naïve selfishness. They are constantly perceiving the significance of things for their own experience, and acting upon this felt meaning. Although not immediately aware of what is passing in their own consciousness, as is common to certain high types of human psychism, yet in their self-interest they certainly know themselves as experiencers. Thus immediate awareness of one's own psychic attitude as such—apperception—is a kind of consciousness of consciousness in measure divorced from consciousness of the object, and so belonging to such a high scope of psychism that it hardly falls within the range of our discussion, which is confined to simple direct emotion—value of things as implying both self and consciousness knowledge. Apperception as a constant reflection and introspection is certainly not original. In its original form consciousness of consciousness is merely implied element in the study of things. The study of conscious self self-possession, self-poise, conscious psychic self-development, is all very late.

Leaving now the general consideration and analysis of self-consciousness in the light of the general doctrine of evolution, let us note how it occurs in consciousness to-day. Let us come to some direct inductive study.

The simplest method and the most direct of studying the rise and nature of self-consciousness is in those experiences in coming to self-consciousness from deep sleep or from coma after severe accident. I say, "I regained consciousness," "I came to consciousness," meaning, not bare consciousness as in mere sensations or perceptions, but a self-consciousness involved therein. In becoming conscious I came to self-consciousness; in becoming aware of the objective, I at once realize my subjectivity, myself as 262experiencer. In coming out from under the influence of chloroform, there is, I have distinctly observed in my own case, a struggling to realize, which is both objective and subjective cognition. It is true a person having awakened under very strange circumstances, as in a bed in a hospital after an accident, may declare, "I did not know myself," but this does not mean that he had no self-consciousness, but merely that for the moment he did not identify this self, himself, as John Smith, of Jonesville, etc. Sometimes it happens that self-identification is not reached at all, but the self, as bodily whole experiencing, is speedily aware of self, a new personality and sense of personality quickly grows up. Again, a lunatic mistaking *himself* for Herod or Cæsar is thus always self-conscious. He has consciously established himself as the self playing a part in the world, but according to the opinion of his sane fellows he is much in error as to what that part is. Strictly speaking, there is no illusion of self-consciousness, except under the impossible supposition that a being not a real self or psychic individual should have self-sense; but the very act of self-cognizance implies reality of self-hood. It is plain that even the insane man who regards *himself* as tree or stone, has, however, the act of self-regard, is really self-conscious. Strictly speaking, we cannot identify or recognise self, for sense of self is necessary in any recognition to make it such, a self-consciousness is a fundamental *prius*. You recognise a tree, a house, but you do not recognise yourself except as yourself is mere object related to you, to your experience. Self-identification means only objective act, and is not, then, the same as self-consciousness, though based upon it.

I have endeavoured to make observations of myself in moments of awaking from sleep or going to sleep, to find whether subjective reference and objective apprehension are mingled co-ordinately in consciousness from the beginning, whether the self-sense reaches through both the 263perceptive life and the sensation life. Drowsing in bed I sometimes have a feeling of bare pleasure as the first stage in a pleasant awakening. There is here no sensing, no localizing, no awareness of body or of anything, no self-consciousness. This mere undifferentiated pleasure, interrupted by "cat-naps," may often recur. Lolling half-awake every one has frequently experienced these feelings of pure pleasure, unsensed and unlocalized, and wholly unobjectivised, the barest and simplest consciousness, the very first stage in awaking. In this very lowest *status* in which I can ever catch my consciousness I have the pleasure from the warmth and softness of the bed without having to feel warm or sensing the soft. It is a distinct step to even feeling warm; moreover, in extreme drowsiness it is an effortful step, an active sensing, an objectifying self-activity, and hence a real self-consciousness, implied in the sensing act. To *feel* warm, to sense in this mode, is primarily object cognition which implies a measure of subject and consciousness cognition in feeling the warmth as source of the pleasure. Any one who will closely examine his mental state at the very first stage of slow awaking from deep sleep—a state of primitive consciousness—will notice a vanishing moment of mere pleasure or pain, and in cases of great drowsiness, when a sensation supervenes upon this stage, it does not merely *come*, as in our ordinary consciousness, but it is *brought*; there is objectifying effort. So in basking in the sun like an animal, the very first and lowest stage of consciousness I drop to is pure pleasure without having even to feel warm; and the feeling warm is distinctly a new and higher step in consciousness which is often attained by some slight effort. Thus it is distinctly possible for a man at times to be too lazy to feel warm; and this fundamental laziness must be accounted not uncommon with lower psychisms. Similarly for cold awakening one. There is a moment of pain from cold before one feels cold, 264a general pain and uneasiness discomfort before one realizes what is the matter, feels cold and the part cold—foot it may be—and so reaches some self-consciousness; in language expression, I am cold or feel cold. Here is a self-conscious personal experience, though the first touch of mere pain was experienced by the individual unconscious of himself.

We infer, then, that self-consciousness is first reached and maintained in the sensing act as definite cognitive volition. To sense warmth and cold is simply a little earlier objectification than to attain sense of a light or a sound. To *feel* is as active as to look or to listen. We know that there are modes of force an appreciation of which does not now enter into known psychosis, but which might be sensed through long and severe effort and evolve a new sense-organ. Thus, if the conditions of life had demanded it, there would have arisen in the struggle of existence a magnetic sense, though now a man may place his head between the poles of the strongest magnet and be unable to reach any sensation. A magnetic sense once organized and inbred into experience would act with the same apparent spontaneity, as a "*given*," as does such a sensation as that of heat; and a person feeling magnetic would have self-feeling implied the same as in feeling warm. That feeling warm with us denotes something which possesses consciousness rather than consciousness by struggle possessing it, is simply the result of the inheritance of the accumulated mental force by which past generations have reached this sense, and thereby consolidated self-consciousness with it, for self-consciousness is built up as reflex cognition from the cognitive effort and willing of the individual. Sensation always begins in a sensing, a volition of the individual to realize externality in its experience value, that is, mode of affection of its own body, as in feeling warm pleasurably or painfully. When the objective is not merely sensed but perceived, 265when object and objects are definitely cognized, self-consciousness is greatly furthered, as each object and objectifying cognizance means self-reference or interpretation in terms of self-experience.

That self-consciousness is early and fundamental psychosis, is apparent, not only from the gradual losing consciousness on going to sleep or in gaining consciousness in waking, but also from the fact of its being universal in dream life. Those factors which remain throughout all stages and kinds of dream life, are justly regarded as organic and basal. The higher and later elements, those which are still nascent and in the volitional stage, as conscience and reason, rarely or never occur in dreams. In the slightest dreams there is personal quality; I am consciously experiencing, I am walking, riding, looking, hearing, etc. An awareness of self pervades all dream life, even in its lowest form. We are constantly in a world of objects which we are conscious of in their experience value as affecting us or to affect us. A person relating a dream always narrates it as personal experience and so felt—"I dreamed I was in a cave and I heard water running and I felt it cold," etc., etc. As far then as we can survey dream life, it is a significant fact that self-consciousness pervades it.

As far then as we can discover in dream consciousness, or in ordinary consciousness, self-consciousness is persistent and pervasive element. In the whole range of consciousness, with the exception of the very evanescent and absolutely primitive pure pleasure-pain series, self-cognition appears. We say, indeed, that a man forgets himself in a rage, but mean merely that the rage object as self-related quite engrosses consciousness to the exclusion of other forms of self-consciousness, as himself related to other selves. Blind with fury to all other objects than the rage object, he does not notice things as related to himself, and he will rush into a stone wall. In the utmost 266concentration and intensification of emotion, self-consciousness does not disappear, but is itself concentrated and intensified. Even in the delirium of passion, so long as any cognition remains self-consciousness remains. The intensification order in consciousness, that is, where multiple consciousness loses elements through intensifying of some others, bears evidence then to the fundamental nature of self-consciousness. A person roused from sleep by cold, which becomes more and more intense till he loses all consciousness through suffering, is throughout the long series self-conscious with the exception of the initial and the final pang of pain. From the moment when cold made him attain consciousness till the moment when he thereby lost consciousness—that is, practically the time he was conscious—he was self-conscious; this is the verdict of common introspection. Any one who looks back upon his experiences of this intensification nature, finds himself to have been self-conscious throughout.

So far then as I have been able to examine them, the modes of coming to consciousness in dream life and in awaking process, and also the order of disappearing consciousness by intensification, confirm the general result which at the opening of this chapter we deduced from a general consideration of psychism under the conditions of existence, namely, that self-consciousness is necessary and important factor in all cognitive process, the self-relating act giving vital value to all consciousness of external and internal object, whether in sensing or perceiving.

We have already touched on the general function of self-consciousness, the gain which accrues to the individual organism from knowing its own experiences as such by giving self-directivity and special response. The individual is thereby enabled to look after its own interests, to consciously care for itself, and to make the most of itself. The core of psychic life is *interest*, and the core of *interest* is self-consciousness. That the psychism has 267interest, that it feels for itself, is essential to the progress of life. Indeed, the genesis and growth of biological forms and organs lie in their attainment and perfecting as servants to the self in the struggle of existence. We know this to be the case for the sense-organs. The organism evidently came to appreciate light by a definite *nisus* with self-consciousness, just the same in kind as that by which organ is advanced to-day when straining the eyes to perceive a seventh Pleiad. In short, we do not see because we have eyes, but we have eyes because we see. The seeing activity and effort as a self-activity generates the eye and perfects it. So also it is by locomotive effort that motor organs originate and develop. The young child learning to walk, self-consciously and with effort moving upon its legs, is an intimation of the way in which the limbs themselves arose in active response to environment. The rabbits imported into Australia have, it is reported, learned to climb trees, with a consequent modification of foot structure. Now the real genesis of the morphological change is obviously psychic, the climbing effort as a valuable function to life under the conditions of existence, viz., the scarcity of herbage.

But not only the motor and sensory organs are to be traced in origin and growth to psychic basis in self-consciousness and struggle, but other organs now quite disassociated from will may originally have been developed by will. Thus the stomach may have originated in digestive effort and the heart in circulatory effort. That self-attention to the heart stimulates the action of the heart is well-known, and also that in rare cases the heart's action is directly controlled by will. This may be survival. Function is built up also as indirect result of will, as when motor effort in running develops heart action. Psychism may thus be interpreted as the basis of all organic development. The body is the offspring of will. Certainly as man surveys progressive adaptation in 268himself and other evolving organisms, the psychic basis is apparent in feeling and in effort self-conscious; and if in any wise it has apparently become mechanical and spontaneous, as in heart-beat, as in digestion, as in winking the eye, this is to be ascribed to impulse from the past. Self-consciousness quickens reaction, for reaction time is shortened when there is anticipation, and anticipation implies self-consciousness as awareness of experienceability. Self-consciousness also enormously strengthens reaction. Thus the more thoroughly one realizes his own danger, the more powerful the effort to escape. This is true under normal and simple conditions, the only form in which we are considering self-consciousness. Self-consciousness may become abnormal and debilitating in the hypochondriac, but this is a stage beyond our present studies. Primarily in the struggle of life self-relating to one's own experience is always advantageous function. The most important thing in life is the realization, by the aid of self-consciousness, of the self-experience value of things; to appreciate and understand environment, and so adapt oneself to it and adapt it to oneself, to conserve and extend self, this is the substance of psychism, and its whole history is thence pervaded by self-consciousness.

But we must now turn from these general considerations to specific emotions as related to self-consciousness. In the natural course of things, an organism can never sense or view the self with indifference. In all early psychic stages a dispassionate view of self is uncalled for and does not exist; and, in fact, even if the most educated and thoughtful human adult had a self-sense which is active as evolutionary cause, it may rightly be regarded as ever active. Life forms from the lowest protista to the highest vertebrate are in their development due to active response, and thus morphological development may be looked at as a functional embodiment of psychism. Instead, then, of regarding psychism merely as life factor, we 269may go farther, and define life as psychism. This is what the doctrine of active response and development thereby, with natural selection, leads to. The phenomena of life, so far as we can interpret them, seem to favour the view that organism is objectification of the will, and, except at the very first stage, will as cognitive, and triply so in object-subject-consciousness cognition. Such evidence as we have points rather to organic body as reflex of mind than mind as reflex of body. That the initiatory, progressive, and creative force in evolution is psychic, we judge from such instances as we can observe of progressive adaptation in ourselves and in lower animals. Where new circumstances affect a species, as the rabbit transferred to Australia, the favouring modification of the foot to climb trees is evidently only attained by severest struggle for self-conservation. If a new mode of force were introduced to this planet, which should powerfully affect life, it would reach it at first only through pleasure-pain, and the growth to a special sense-organ for this new force would very gradually be attained through the struggle for existence.

The prime value of self-consciousness in evolution is in securing an intelligent correlation with environment. All specific reaction and adaptation arose probably through an emotion volitional self-relating of object. It is a biologic psychic law that all emotion is bound up with self-consciousness, and all self-consciousness with emotion, for thus only is there efficiency as intelligent will stimulation. But while sense of self is inherent in all emotion as such, may it not in some cases have a peculiar place, so that we may justly term them self-feelings or emotions of personality?

A child fears the dog and is proud of its new dress. Here are two emotions which both imply self-consciousness, the object is in both related to the self, but they differ in egoistic quality in that in the fear there is sense of the thing as acting on the self, in the pride there is 270sense of the self as acting on the thing. In the pride it is the object as identified with the self that is the source of emotion. The pride proceeds from within outward, while fear, *vice versâ*. In fear it is the experience value of the dog, that it will hurt, that gives the emotion quality; but in the pride the essence of the emotion lies, not in the influence of the dress on the self, but that the self is connected with the dress by way of ownership. "See my pretty dress"; "Oh mama! the cross dog"; the emotions thus expressed appear to belong to different orders; the fear being of the thing in its effect on the self, the pride being of the self in the thing. Pride is a glorified self-consciousness, self-consciousness is its substance and immediate spirit, whereas in fear self-consciousness is but an instrument in interpretation of experience value. We observe an interesting example of emotion of personality in a young girl who fears a cow and is yet ashamed of her fear. Here, while self-consciousness is certainly involved in the fear, yet it is peculiarly involved in the emotion at her emotion as such; the shame is at or of herself, the fear is for herself. This peculiar personal feature of pride is signified by the common usage of language; the child is proud of the thing, does not pride the thing, but prides himself on the thing, whereas in fear he fears the thing for himself. I say, indeed, the child is afraid of the dog and proud of his dress, but the force of the preposition is quite general.

It may be said that pride is not peculiarly an emotion of personality simply as being directed toward self; one can hate himself, fear himself, be angry at himself, etc. But the drunkard fearing himself means merely that he fears the results of his own tendencies, *delirium tremens*, for instance, a perfectly objective fear. And it is evident that one cannot, holding to the term, self, in the same meaning, fear at once himself for himself. The self 271which is endangered is not the self which endangers. In all such cases as so-called fearing self the action is from without inward, which is the reverse of the mode in personality—emotion where oneself is seen, not as affected by the thing, but as himself in the thing.

The typical and earliest of the emotions of personality is undoubtedly pride. Like all emotions pride includes cognition of object; pride is always proud of something but in the peculiar way before emphasized, in the light which our own personality casts upon it. Pride generally and certainly originally implies sense of something done or possessed by self and that in a manner superior to competitors. It is a self assertion over rivals, an impressing spectators, a being proud of something to some one. If the world contained but one solitary conscious individual, he could never attain to pride, though he might be self-satisfied. Sense of comparative self-magnification is essential to pride. Pride as social in its nature suffers great diminishing when the individual is long kept in solitude, and in some cases men may ultimately lose all standard of comparison and so pride entirely vanishes. If a man were from his earliest remembrance an inhabitant of a desert isle pride would have no opportunity to develop. His achievements might satisfy himself, but they could not make him proud, for he would know nothing of others and their works. Again, this need of sociality is seen in this, that we are not proud of our planet as such. We distinguish it, indeed, as our own, but we have no sense of pride in its finest features as such. I do not feel proud of Amazonian forest or Himalayan mountain merely as earth characters. However, if in the future we secure interplanetary communication, and planets rival each other as cities and countries do now, there will be a stimulus to pride on an astronomical scale. If we could say to the inhabitants of some neighbour sphere that our planet made better time 272round the sun than theirs, this would be the basis of an intense pride.

The extent of pride is thus equal to the extent of the self-sense, but in its wide ranges pride is relatively weak. I am proud of my country, but, other things being equal, more proud of my state and still more proud of my city. I am proud of the achievements of the Anglo-American race, and I always survey a locomotive with pride, but it is when ownership and achievement comes closer to the *ego*, as in one's relatives and family, that pride notably intensifies, and it reaches its *maximum* in view of one's own attainments. That which we do without any assistance and which seems to us far beyond the ordinary gives the best and highest incitement to pride.

Pride, in the later stages at least, is more and more discriminating, and is connected finally only with those objects which are the actual will products of the individual, and so identified with the veritable self. Thus is erected by society a pride test, and men say, "He has a right to be proud," or, "He ought not to be proud." Yet standards will differ, and what one will be proud of another will be ashamed of, and *vice versâ*. The general standard is largely regulated by the comparative amount of will force and so of strength required in the particular act; thus, while I am not proud of crushing an ant, I might be at felling an ox.

The general expression of pride is holding up one's head and expanding oneself generally, though this self-enlargement is not, as in anger, to inspire fear in beholders, but rather admiration. Proud sense of superiority naturally asserts itself primarily in physical impressiveness, and, as such, pride plays an especially large part in sexual selection. The lower expression of pride is swagger and strut, the higher in a dignity and stateliness of demeanour.

273The function of pride, the use which originally determined its development, and which is still apparent, is a pleasure-sanction to competitive successful effort. The proud consciousness of triumph is one of the greatest pleasures of existence, and if there were no such emotion following the winning effort, life would lose much of its incentive. Pride prevents parasitism. Without pride to stimulate and reward, striving mind would have lost one of the most potent factors of progress. Even in human education it becomes of value to appeal to a just and proper pride. In the lower life it is all important. It gives tone to life, gives power and confidence, assertiveness and aggressiveness, and conduces in a large measure to permanent and progressive self-aggrandisement. And not only for effect upon self but upon others, pride is an important psychic factor. Thus pride in always showing a bold, commanding front to rivals, makes a direct impression upon antagonists. Pride always puts the best foot first, hides weakness and exaggerates strength, so that the proud one always shows for all and even more than he is, and thus gains much in the struggle of existence where even mere appearance of power is apt to discourage opponents. The one who is strong and proud of it is doubly strong. Pride is the reflex of gain and victory, as shame is of loss and defeat. It is thus the root of ambition, the desire of rank and place for superiority's sake which has been, and now is, especially in advanced human psychism, a most powerful agent in the evolution of life and mind.

But while it is undoubtedly true that pride is in its origin solely an advantageous psychosis, and indeed, could have been developed in no other way, yet there is a disadvantageous side. Only up to a certain point is it true that the prouder one is, the better off he is. When pride, over-stimulated, betrays into over-confidence and heedlessness, then, indeed, "pride goeth before a fall." 274But at the first, however, we must suppose that the organism was proud of only that of which it was to its advantage to be proud; but by perversion and hypertrophy, indeed, in pride as in the case of other emotions, caused largely by rivals, it became a source of great disadvantage and positively destructive of high self-advancement. Conceit, an over-weening abnormal pride which is totally irrelevant to the real standing of the individual, cannot but be highly injurious. However, harmful pride must be accounted rather late. In early psychisms attainment over and beyond others, when perceived naturally and normally, gave rise to pride as a wholly useful emotion reaction, and those who had the capacity of being proud had a distinct advantage over those who had no sense of their own consequence or no pride about it. Even in human society we must remark that in general those who are incapable of becoming proud on proper occasion, are less and less liable to reach the occasion.

Pride, as emotion of sense of superiority, manifests itself in many forms, of which we need not now expect to make a detailed or complete investigation, since the object of our present studies is merely to emphasize the main forms of the early emotions from the point of view of natural selection. Simple pride, which is unconscious of itself, but acts directly and without reflection, as we see in a child proud of a new dress, is a phase which does not often appear in the experience of the educated human adult, where pride becomes highly complicated with emotional and intellectual movements of many kinds, and where it is extended to a wide diversity of objects with the extension of self-interest. Thus men are proud of rank, blood, money, muscular strength, possessions, intellectual attainments, moral character, and, in fact, whatever the idea of *mine* can be applied to. However, the different kinds of pride are 275not to be distinguished by the object merely, as pride of rank, blood, etc., for difference in object does not by itself constitute distinct quality in psychic act. Pride is the same, whether it is of a horse, a bank account, or a wife. Still the object frequently calls up subsidiary emotions which may complicate pride, and the perceived nature of the object certainly influences our feeling toward it.

When an object is to be competed for, but we consider it beneath us to enter the lists, or we think our rivals unworthy of our attention, we have the peculiar phase of feeling termed arrogance. Arrogance brooks no rivalry and stands apart on a peak of self-contained superiority. Walter Savage Landor, the proudest of men, displays this feeling in perfection when he says in one of his cameos in verse:

"I strove with none,
For none were worth my strife."

This is a perfect expression of complete arrogance. We may say that he was too proud to be proud. No one was worthy of his mettle, and so he held himself aloof with the feeling of immeasurable superiority. Strictly speaking then, arrogance is a variety of very intense pride where the sense of superiority is perfectly exclusive and absolute, and disdains comparison. It is entirely inconsiderate of others' rivalry and above caring for the approval or disapproval or admiration of others. Thus this phase, unlike pride in general, seeks concealment rather than display; its excellence is so far beyond the common as to be unappreciable by contemporaries, and appreciated by self alone.

Conceit is a term objectively applied, but hardly indicates a kind of pride, a real subjective distinction. He who thinks more highly of himself than he ought to think, esteems himself beyond his due, and so is considered by the community over proud, is termed 276conceited. The pride which is entirely just, as viewed from the objective standpoint, is quite the same subjectively as the most preposterous conceit. Similarly also dignity is no real feeling. "That man is dignified"; this is an objective characterization of his manner of conduct, but this does not imply that he feels dignified. Pride may give a dignified demeanour, but a feeling dignified can only refer to the reactive effect upon consciousness, of this mode of behaviour. "I feel proud," may likewise sometimes be used, not for designating the subjective feeling or being proud, but as equal to, "I felt that I was proud," that is, "I was proud and I knew it," "I had the sense of being proud." So also in general we may remark that while feeling may denote a simple state of being, yet such phrases as "I felt proud," "felt angry," etc., are ambiguous, and may mean either the bare feeling of pride, anger, etc., as experienced, or the feeling of being proud, angry, etc., or both, that is, consciousness of the particular consciousness may or may not complicate self-consciousness. The word, feel, is often used in this merely reflexive way to denote a sense of state as, "I was proud and I felt so at the time." Thus common phrase verifies the analysis that self-consciousness and consciousness of consciousness are bound up with emotion, the full analysis of the phrase showing that the feeling proud was an object consciousness *plus* a subject consciousness.

As previously intimated, we have to sharply distinguish between pride and such emotions as self-satisfaction and self-complacency. These latter emotions of personality deal solely with the self in its own sight, while pride is always not over self to self, but over self to others. The self-satisfied often are proud, but this is not necessarily implied. The comparative element enters in self-satisfaction, as in all true pride, but the comparison is primarily with oneself, not with others. If we succeed in our own 277eyes, we may think little about others. A pure self-satisfaction, like a purely altruistic pride, is a rare and late phenomenon. Pride about others, pride to oneself, are both very apt to be tinged with the original pride over others. One says of a friend, "I feel proud of him"; but while this has a certain reality and psychic value of altruistic mode, yet the innate and fundamental selfishness of pride tends to make a place in what appears to be the most disinterested form. Personal interest and aggrandisement is so inbred a motive from the earliest stages of evolution that it is never superseded.

A feeling of embarrassment is an emotion of personality which is closely connected with pride. Those who are most susceptible to pride are most apt to feel embarrassed. The one who has no tendency toward pride, who does not in the least care how he may appear before others or in relation to others, and so does not value his place among his fellows, cannot be embarrassed. He may be disturbed by the difficulties of some task, but only in the same way in which he would be agitated by any difficult work undertaken by and for himself alone. The emotion of embarrassment, like pride, conceives the self in its social relations. When one says that he felt greatly embarrassed in being called on unexpectedly to speak at a dinner, we perceive that he means emotion, not merely in view of the inherent difficulty of the task, but in view of what he himself may or may not do under the inspection of the critical. In this emotion there is a wonderful quickening of the self-sense, a painfully intense self-consciousness being suddenly generated as the peculiar relation of self to others is impressed upon him. This self-sense is powerfully reinforced by the self-sense of the bodily expression of self-consciousness. The whole bodily self seems conspicuously magnified, and we become painfully aware of hands, feet, and other members. This bodily self-sensitiveness, as often contributing strongly to this emotion-total, is very 278marked in cases of blushing. A girl, feeling embarrassed, blushes, and immediately becoming conscious of the blushing as itself an embarrassing circumstance, blushes again still more violently, and becoming conscious of this, becomes still more confused, and so on, a constant cumulation of psychic effect from reaction of expression. Sense of the expression of embarrassment is itself embarrassing, hence every embarrassment may become in itself a new source of embarrassment. However, that this peculiar self-consciousness cannot be forced in itself or in its expression, we see in the fact that the efforts of the maiden who exclaims in mock modesty, "I *know* I am blushing," are entirely futile. This assumption of embarrassment may become embarrassing, and so a genuine expression be stimulated, which, however, is of quite another order from the one desired.

How such an emotion as that of embarrassment, which is disadvantageous from the first, could have originated under natural selection, can never be solved by the evolutionist who views all variation as originally springing from personal advantage. Here is a psychosis, always the reverse of serviceable, an emotion anticipatory of disgraceful defeat, and so is really premonitory, but yet one which ever unnerves, rather than nerves to successful action. He who never feels embarrassed, under any circumstances always has the best chance. Hence this psychosis must be strictly a negative evolution, an unfavourable variation determined by a persistent exciting by antagonists as serviceable to them. An adversary will always put his opponent in an embarrassing situation, and endeavour that he shall both be embarrassed and feel embarrassment. This emotion has thus been stimulated and fostered during ages of psychic evolution, and in advanced human evolution the stimulating it is one of the subtlest methods of offence.

A feeling of embarrassment is incipient shame, or perhaps 279the way for shame. But the feeling of embarrassment is generally anticipatory as to the potential, while shame is as to the actual; it is a feeling of present public degradation and loss. Both equally imply a capacity for pride; one who cannot be proud cannot be ashamed. But shame, unlike the feeling of embarrassment, acts as serviceable variation to the individual, and is one of the weightiest negative guards to advantageous actions. It cannot promote very high and noble action, but it keeps above a certain low and base level. The member of society who has lost all pride and all sense of shame has ceased to feel the most powerful and useful of social incentives.[E]

E. As to the origin of bodily shame, we may suppose that this arose with reference to *excreta* as something rejected from the body, and therefore base and unworthy. With the refined even spitting and perspiring are shameful. It may be that sexual shame can be traced to the same root, but social convention and morality also have very large influence here.

There is a certain curious psychosis which may be called shame for want of a better term. I allude to the feeling which prompts one to shun oneself. One may not only be ashamed to look others in the eye, but even himself. He will not look at himself in a mirror because he feels a great loss of self-respect. This is not the opposite of vanity, a shame at viewing oneself because of unseemliness of feature, which is liable to general observation, but it is rather the reverse, the polar opposite of pure self-feeling, of self-respect and self-satisfaction. A feeling of shame with regard to oneself alone is still, of course, comparative; though it does not touch upon others, it implies a self-erected standard. This emotion, like the others just mentioned, is obviously very late.

However, perhaps the latest in the series, and the psychic culmination of all is humility. Humility, like meekness, marks a new order of evolution. In the highest human development pride is eliminated and supplanted 280by humility. A true self-estimate of personal achievement upon a very wide and impartial impersonal basis, either that of a scientific view of man's place in the universe, or as influenced by high religious and moral ideals, leads to a feeling of humility. Egoism and self-assertiveness give place to altruistic modesty and refined reserve. The humble man always gives place rather than takes place. He does not lift himself above his fellows, but takes the lowest seat, and is servant of all. The humble man does not strive with others, not because too proud to do so, as Landor, but because he feels called to the highest and best work for its own sake. He says with Laotze, "Do, not strive." Unthinking of getting ahead or falling behind others, he aims consistently and constantly at an ideal of perfect fruitage, so high an ideal that he always feels his own unworthiness in his own sight and in that of others, though aware of his desert by the ordinary standards of his community, country, or generation. Worldly successes produce no elation in the lowly of heart; they view themselves, not with self-depreciation, but with the justness of the largest view, as Newton, who, when complimented upon his attainments, replied that he had but picked up a few pebbles by the ocean of truth. Spiritual and ethical principles sway these, and not personal ambition. And it must be noted that humility is not simply lack of pride under circumstances which naturally allow of it, an insensitiveness to pride, a wholly negative state, which is nothing in itself, but it is a positive feeling and emotion in view of oneself in relation to others. Thus the humble man is he of high pride capacity, and who consciously refrains from pride when usual standards would allow it. "That is something to be proud of," "He has a right to be proud," and similar expressions mark the lower standards of which he never avails himself. The best and noblest specimens of mankind renounce the "world," "the lust of the eye and pride of life," and live by their self-erected 281 ideals. And if we ask how the spirit of humility and disinterestedness can arise and progress in a natural evolution, we must answer that it holds its place and wins its way by reason of its greater inherent value and fruitfulness. He who has himself in view has lost sight of his work. By this psychic mode alone is the largest, most perfect, most permanent work accomplished, and ultimately, often posthumously, it is appreciated at its real worth. Those originating and master minds in human history who have opened new avenues of spiritual progress, have usually been of this modest, unassuming, humble type. Thus in a wholly natural manner the higher law of an ideal life prevails over the lower law of life which works only by competition in the struggle for existence.

CHAPTER XVI

INDUCTION AND EMOTION

We have implied throughout that we have feeling about a thing only so far as we attach on basis of past experience an experience value to the thing, as we say, "the burnt child dreads the fire." Induction, as this interpretation is termed, is so important an element that we will devote a little space to considering its *rationale*, development, and place in intellectual emotion.

What is the *rationale* of the inductive act? Why should iteration lead to expectancy of reiteration? I observe that a body unsupported falls in a hundred instances, but is it not arbitrary for me then to suppose that it will fall the hundred and first instance? In fact would it not be more rational to suppose that this particular combination should be exhausted, that it was time for nature to stop? But this very reason rests on the uniformity of nature—the very law we are questioning—as experienced in the past and applied to the future; only it is a negative law of omissions, literally law of reiteration of unreiterations. Thus if reason takes the law of uniformity of nature to task it can only do so by assuming it. J. S. Mill in his treatment of this matter (*Logic*, bk. iii. chap. 3, sec. 2), falls into an error. It is, indeed, true, as he says, that some occurrences repeated suggest cessation and not recurrence, as when we have several consecutive cloudy days, we expect a bright one, or having had several rainy seasons we expect a dry one; but it is plainly wrong to 283regard this, as he does, as a contradiction of the principle of uniformity of nature. On the contrary, this is a very good example of it. Experience of intermittent character of bad weather in the past leads to expectancy of its re-intermittency for the future, and the oftener the experience the stronger the belief as to the nature of the still unexperienced. A negative uniformity is as much a uniformity as a positive.

It is plain that we can assign no reason for our belief in the uniformity of nature. It is simply a fact, an arbitrary fact if you will, that the more often experiences are conjoined, the more strongly we expect the conjuncture. I may imagine a body unsupported remaining stationary in the air as readily as to imagine it falling; however, I *believe* it will fall, and I duck my head for fear of getting hurt. Not any speculative reason then, but a very practical reason, is at the bottom of this inductive tendency, that is, the conservation and progress of the organism is secured by induction as anticipatory function. The origin of induction is not then in its abstract rationality, but in its immediate utility as a life function. Experience is self-adjustment through felt stimulus. Once begun it grows by continual self-reference, and hence practically all experience is inductive. Experience is thus a *continuum*, an integrating cumulating whole; and inductive experience, like all experience, arises and progresses by reason of its serviceability.

It has been implied that the inductive act arises very early in the history of experience. Every psychosis is what it is by reason of all the previous psychoses in the individual and the race. Psychism, while it has its points of development in individuals, must yet be estimated as a unit, as a single whole. But we have to ask whether this modification of one psychosis by another is conscious or unconscious. If some low organism have in its lifetime but two consciousnesses, must we regard the second as 284influenced in quality by the first, and if so, consciously influenced, that is, a conscious relating, an active induction as opposed to mechanical integration? Is mind always self-building, or does psychosis act and react on psychosis automatically? We have maintained that all the growth of mind has been in the past, as in the present, by struggle, by severest endeavour, and hence if experience modify experience it is by conscious act. Experience thus constantly connects with itself and builds upon itself, it is self-integrating, that is, inductive, in all its evolution. Mind, as primarily pleasure-pain and struggle, by endeavour reaches back to itself, realizes itself, and rises upon itself.

Take a comparatively simple case. A child tastes an orange, and finds it sweet, *i.e.*, it relates the sweetness to the object, which relating is a true thinking, an active conjoining or associating. Upon the presentation of another orange to the child at a later date, he identifies it as the sweet thing; he associates sweetness as to be experienced from it on basis of past associating, that is, he makes an induction. In this second orange-experience, as far as there is active conjoining of mental products, a definite adding to present percept of sweet taste as experienceable by conscious reference to former percept (taste-experience), we must recognise a genuine thought-process. The thinking consists in the joining of sensation of taste to an object, not as a present, but as a future experience, on the basis of some past experience. Here is a true mediation or reasoning of inductive type, and also a true concept-process, that is, a taking together, a conscious uniting, although the product is still particular. The nearest approach to expressing this psychological process in language is to say, "This round yellow is this sweet, because this round yellow was this sweet before." The correlating process rests upon the relating process accomplished at first experience of orange-tasting, whereby 285the taste was related to the thing tasted. This relating may be thrust upon the mind, or the mind may consciously and actively assimilate. Thought in the wide sense of the term may be made to include all mediate or immediate conscious conjoining of experiences, whether the product be general or particular.

Mediacy is certainly, however, accomplished before commonness is noted, which in ordinary usage is concept-making. The grouping of the particular taste with the particular sight and touch on basis of past experience does not give a general result. The mediate term of past experience of taste which the child brings up on sight of orange and applies to the present case does not suggest commonness, but constancy of experience, for at first it knows things only as identical, and not as separate, or as like or unlike. The method of this early intelligence is that of identifying, "The orange was sweet and is sweet"; and not that of common characterizing, "Oranges are sweet, and this is an orange." The child does not discriminate or understand that the object of its first experience is, by reason of this experience, no longer to be experienced; it has not attained notion of disappearance. It does not cognize the orange as one of a group or class, having as common characters roundness, sweetness, and yellowness, and from presence of round-yellow in any instance infer sweet; but it knows orange only as this particular object of past, present, and future experience. Many of the early thought-experiences of children are to be interpreted rather upon this identity-method than upon the usual interpretation of true concepts. Thus the child who calls every person of certain age, dress, etc., "Papa," is not thinking of a papa, or class of papas, but of the papa. This is mistaken identity: the common and like is the same, and the child requires considerable discrimination before it attains to notion of papa in general. Same and not-same are discriminated before like and unlike, and 286hence young children use common names as proper. Now the mental product achieved by the child, which, as expressed in words, we term the papa, may be styled a particular concept, a gathering together of sight-sensations, and associating sound- and touch-sensations with these so that any generally like group of sight-sensations enables the child to call up on basis of past experience the associated sound and touch, to expect the gentle word and caress. The child in identifying the orange, "This round yellow thing is the sweet thing," is bringing together with a certain general force, not of common characterization, indeed, but of temporal significance as permanent grouping. Animals and young children think mostly on the identifying plan; they join to and expect for a present experience what has been conjoined with it in past experience, but the object is the same, not a like one.

How then does the child come to knowledge of things as like, to form a class of oranges after regarding all oranges as the orange? Pass oranges before a young child one after the other so that one only is in sight, and the child will probably know only one orange as the same continually re-appearing. The image formed will, however, be more or less composite, the mental product will be a concept-image, as being a re-inforcement and exaggeration of common characters and a suppression of individual; but for practical purposes it is still a particular concept, that is, the child applies it to the one and not the many, and does not recognise its representative nature. A general image as a group of common qualities may be thus attained before consciousness of this generality is reached.

If now two or three oranges are presented to the child at the same time, it will learn to discriminate them as separate co-existences, having characters in common, roundness, yellowness, etc.; the objects will be recognised as individuals belonging to class round-yellow things. Here a general image having a general import is achieved. 287The particular characters, round, yellow, sweet, which always centred in and made up the individual orange, are recognised to have general scope in applying to many objects. Groups of characters had been achieved before by particular thinking, but now by general thought groups of characters as common are formed. From the practically coincident impressions it gains the notion orange, so that it recognises new individuals as individuals, and not as the individual or single object, as in the earlier and cruder identity method of thinking. The mind now—instead of saying "Same impressions, same object"—says "Same impressions, like objects." Instead of making an object as a group of qualities, it makes a class of objects having the group of qualities in common. Concept-forming is thus often but an extension from what I have termed the particular concept; the group of qualities formed as characterizing the thing is through experience with co-existences predicated of things. Notion or idea of the orange precedes notion or idea of orange; but both are truly notions or concepts, a taking together of impressions, one of particular, the other of general import. The general significance of the particular group is first forced upon the mind by experience, but soon the mind generalizes as well as notices generalizations brought to it. Gradually the mind obtains power to generalize, not only from co-existences, but from successions, and later still to generalize by abstraction, to compare and pick out common features amidst the unlike, to search for unity in diversity.

The rise of generalizing power is through the struggle for existence; it originates, like all other mental processes, in practical needs. Law is thereby not simply acted upon or merely recognised, as in the associative stage: it is definitely sought for and applied. Art arises, and also science. The ability, given by generalizing power, of dealing with things in the lump, becomes of signal service, and specially distinguishes man. But the primary value of the 288concept in all its stages is not as a summation of experience, but as a guide for the future. Through reiterated grouping the concept-group is recognised as permanent factor, so that one element of a group being given, other elements are expected through a conscious assimilation with the past experience. The concept answering to the word orange, for example, is the mental product recognising a constant co-existence of certain qualities of shape, colour, size, taste, etc., so that from occurrence of one or more we infer other or others. Concepts are the inner groupings, the mental synthesizings, which interpret the outer groupings that we term laws of nature. In all this we see the inductive element in its conscious form, experience developing itself by anticipating future in terms of past.

We have now to consider briefly the psychological nature of judgment and reasoning with special reference to the inductive feature. Logically judgment is any connecting, *plus* affirming of reality, as effected through the copula. The copula is made, not only to denote relation, but reality of relation, to express, not only the act of connecting, but also its validity for the case in hand. Psychologically, judging may be regarded as any thinking, as any relating without reference to the things related, whether it be a joining of the concept "reality" to some other concept as a concept-forming process, or any joining of other elements. I have already discussed the nature of relating *per se*, but on the topic of judgment a word is to be said about the proposition-form. In all thinking there are the two things joined—subject and predicate in language-expression—and the act of joining, or copula in language-expression; thus all thought is capable of the proposition-form. Indeed, the word-form cannot express a thinking but only a thought as a consolidated and single product, and as a sign of process. The word is a summary of process and relations, but it cannot express process as concept-forming 289or judging. The word orange signifies for the mind by symbolic and shorthand method, "Thing is sweet *plus* thing is yellow," etc; but as far as process happens, and not simultaneous composite representation, the process is capable of proposition-form. All relatings or joinings, even of particulars to particulars, are of the proposition-type, and I must dissent from the common view that two percepts cannot stand in subject-predicate relations. As I have before discussed, the relating of particular to particular is thinking, and to say "This sweet belongs to this yellow" is awkward indeed, but still psychologically proper. Every proposition, on the other hand, is susceptible of analysis as expressive of concept-forming relating. The proposition "Man is mortal" is expression of a mental process of joining; the concept mortal is either attached to or detached from the concept man, according as we consider the process as synthetic or analytic. If it be a grouping or concept-forming in full sense, it means that in forming the concept man, we add to the already gathered qualities the quality "mortal" on basis of experience. The child first notices deaths in cases of John, Peter, etc., whom it knows to belong to the class "men," forms the concept "mortal" and adds it by generalisation to the whole class and enlarges concept "man" by one quality. This proposition, as denoting inductive concept-forming, expresses the act of incorporating on basis of experience the quality mortal into the quality-group man. As analytic, as a detaching of what has been grouped, the proposition still expresses joining, and until the statement becomes purely formal and practically meaningless the rejoining is always a strengthening of the concept, and formative in its value.

All uniting or relating is, however, more than a bare connecting; it is a definite mode of relating, it has a form; and the first and fundamental form is that of time and space, by which all relating has the inductive quality of 290relying upon the past for the interpretation of the future. But thought as self-active mentality is specially stimulated and controlled by the form of reality. All relatings are not, however, influenced by sense of reality, and hence belief is not coincident with judgment in the large sense. Affirmation or denial of actuality or reality is a kind of joining, but is not joining *per se*. The infant joins taste of sweetness with percept round-yellow for the first time and for many following times with no reference to reality or unreality. "This round-yellow is this sweet" expresses a mere connecting, a bare relating, but as neither real nor unreal. There is no emphasis laid on the copula by which it expresses more than a mere joining. But let the perfect tranquility of the child's experience be broken in upon by discord of appearance and reality, let the child once have a bitter experience with a round lemon, then its future conjoinings of round-yellow and sweet will be more or less tinged by sense of possibility of error, and emphasis will be laid on the copula, "That *is* sweet." Through other such experiences with other of its thought-groups, the child generalises to the universal significance of reality and unreality for all its thinking; hence, all conjoinings with their copula-expressions attain a new force and quality from this induction. In the light of fallibility as making up a part of the concept "experience," all thought-experience modifies itself by this self-relation. Reality becomes so constant and universal for all thought-life that mature thought can never escape it. Hegel tried to rise superior to the notion of existence, but psychologically, at least, he failed. The conception or induction of reality becomes a necessary form of thought by being united with all unitings. Judgment in the narrow sense may be defined as all those relatings in which the reality of the relation is affirmed or denied.

Lastly a word on the nature of reasoning. Reasoning is mediatorial; the joining is accomplished through one or 291more mediates. Most if not all thinking is by mediating; joining proceeds only upon ground or basis, whether recognised or not as such. "Kings are mortal" is the language-expression of a conjoining effected either through the particular mediate term, John, or terms, John, Peter, etc., or through the general mediate term men. In both cases the conjoining is effected through subjoining of the mediate term to both the elements to be conjoined. In the first case the process is: "John is mortal, John is king, therefore kings are mortal." This relating of king and mortal is strengthened by subjoining for other particular mediates, Peter, James, etc. In the second case the process is: "Men are mortal, kings are men, therefore kings are mortal." In both cases the appeal is to constancy of coherence of a quality to a quality-group, in the first, mortality coherent with king John, hence coherent with kings; in the second, kings have mortality because mortality is coherent with the group man = kings + others. In both cases the generalising tendency, that is, the inductive quality, is the main point and not the method of mediation. In both processes the concept king is filled out by the additional quality mortality, and there is real gain in generalising and concept-forming, whether the mind accomplishes it by the more special or more general reference. Induction in the large sense is thus inclusive of both induction and deduction in the restricted sense as determined by the mode of mediation. Inductive thinking, as we have treated it, is the joining which generalises, whatever be the means used to this end. Induction as generalising tendency is imbedded in experience, and is the largest factor in all its development. All cognition as interpretation is induction.

That induction, as giving the experience value of things on basis of previous experience, is fundamental to all emotions about the things, has been implied throughout our discussion. But the inductive act may itself be 292emotively considered and intellectual emotion may arise. How and why induction came to be a pleasurable act and carried on for its own sake is, perhaps, not explainable by biologic evolution. It is certain that the inductive act, like other functions, arises as painful effort and as a mere means of serving life. Identifying and recognising is accomplished only under pressure of the struggle for existence. Animals in general, and, indeed, most human beings exercise their intelligence, make inductions, only as compelled by the demands of life. The Australian savages who guided Lumholtz in his search for new marsupials knew about the animals solely in a practical way, and were totally unable to comprehend Lumholtz's motive. So geologists examining stones are entirely misapprehended by savages and the semi-civilized, though these people are sufficiently acquainted with stones so far as they are a source of mineral wealth, are useful for building, etc. And from the point of view of natural selection pure science, the pursuit of knowledge solely for its own sake, without the least reference to its appreciation, is unexplainable.

Nevertheless, it is a fact that at a certain point in psychism the intellectual life develops for its own sake; the inductive act is pleasurable, and the desire arises to continue it as such, that is, here is true intellectual emotion, an emotion arising about an intellectual act represented as such. This feeling about induction may rise to an absorbing passion, as with Charles Darwin. He liked nothing better than making inductions, until he finally came to like little else. If the reason is asked for induction becoming pleasurable, and an end in itself psychology at present has no answer.

It is plain that when intellectual activity is desired, not as a means, but as an end in itself, it excludes much intellectual emotion which is commonly associated therewith. Surprise and wonder, for instance, are intellectual 293emotions at contemplating a conjuncture very contrary to expectation, entirely opposite to some pre-formed induction, but they do not imply devotion to intellectual activity as such. The visitor to a biological laboratory who, on first seeing blood corpuscles, cries "Wonderful! Who would have thought it! One's blood all full of such things! Let me look again!" is hardly actuated by the scientific motive. All phenomena are equally wonderful or wonderless—which amounts to the same thing—to the scientist; for him everything is simply natural, he forms no expectations not founded on facts. The "wonders of science" are wonders only to the outsider: the scientist takes them as matter of fact. It is no more wonderful to him that the blood should be full of corpuscles than that it should fall in drops. The tyro does not wonder at a drop of blood; he wonders to see the drop filled with myriads of animated corpuscles; the scientist wonders at neither. He who, on being told of blood corpuscles, exclaims, "I want to know," plainly desires knowledge, but is not impelled by a pure thirst for knowledge. The scientific items appearing in the newspapers generally appeal merely to seekers for marvels and lovers of intellectual sensation. Surprise and wonder are then extraneous impulses to knowledge, the impulse to knowledge for its own sake being quite distinct. As based on intellectual shock they imply a considerable intellectual integration, and hence are by no means primitive in mental life, yet far from being as late as emotion for knowledge *per se*. Wonder gives birth to the Arabian Nights and to Jules Verne's romances, but it always hinders true science.

Again, the pleasure and desire of achieving and achievement often plays a large part in intellectual pursuits, as in a wide variety of activity. Reaching an end merely for the sake of accomplishment, an emotion about any end, as, for instance, a wide generalisation to be attained, merely as end, intellectual action has in common with all 294other teleological action, but the teleologic emotion is not distinctly intellectual. The desire to achieve for achievement's sake, to reach the satisfaction of accomplishment, is extremely multiplex in its application. The man who does a thing just to see if he can do it, who does feats of any kind is obviously impelled by a different emotion from the one who performs the same activity for the pleasure of the activity itself. He who plays a game to succeed, and he who plays for the pleasurable activity involved, are in very different frames of mind. And emotion for achievement is generally complicated by desire to be thereby superior to one's fellows. The intense competitive struggle is plain in all departments even among scientists. The emotion of competition, the earnest desire to surpass others in interpreting nature and life is a tremendous force among all scientific workers, and not even Darwin himself, exceptional though he was, could keep out every vestige of *amour propre*.

We note also that love of any intellectual activity for its own sake, as induction, must be distinguished from the love of truth. Here induction is exercised, not for itself, but as a means to an end, truth; inducing is not merely a pleasing exercise, but a means to accomplishment of a definite result. Darwin, of course, a trained and habitual inductionist, worked both from the pleasurability of the activity and from his devotion to truth, to which this induction was the true method. Though both these motives, love of an activity and love of some definite end thereby attained, as truth, reputation, etc., are closely connected, they are perfectly distinct modes of emotion, as the least reflection convinces. Truth is some very wide permanent and significant conjuncture of experience discovered and set forth, such as the origin of species in progressive modification, or the intensity of light in inverse proportion to the square of the distance, and this is the kind of induction or conjoining demanded by the love of truth.

CHAPTER XVII

THE ÆSTHETIC PSYCHOSIS

The problem of the origin and nature of æsthetic feeling is a definite psychological problem to be solved only by introspection careful and prolonged. We must take simple cases and closely scrutinize them to discover the distinctive quality, we must seek the cognitive, feeling, will elements, we must note its kinship to other psychoses, we must endeavour to analyse and determine whether it be simple or complex. Analysis, indeed, as chemical analysis, *e.g.*, is a reducing the manifold to a comparatively few elements, from which by composition an indefinite number of substances are formed. But in psychological study we must proceed without any bias from physical investigation. We cannot reduce mind to the mechanical development of a few simples as we survey the development of matter chemically. If mind be essentially self-activity, will effort, then conjunction of psychoses is due to a conjoining activity, and is not mere aggregation. So in case of fear we found a great complexity of conditions, yet fear in itself seems an unanalyzable emotion wave. In taking up æsthetic psychosis we attempt an unbiassed introspective study.

The æsthetic psychosis has been by many evolutionists connected with sexual appetite and emotion. The evidence for this is that among animals the brilliant-hued, and, as we term them, beautiful mates are chosen in pairing time. Also graceful movements and melodious 296tones are then employed. In mankind the æsthetic feeling, as every one may recall in his own case, arose, and became prominent when near or in the teens. The rude boy and the hoyden girl then dress and adorn themselves, and a glamour of beauty is thrown about one who was once an entirely indifferent object. All the surroundings, artificial and natural, of the beloved object are looked upon and thought about in a new way of feeling, an air of attractiveness and beauty envelops all. The period of life of strongest sexuality, from twenty to forty, is also the period of strongest æsthetic emotion. Further, sexuality is notedly strong among those who professionally cultivate the æsthetic psychosis, as artists, musicians, and poets: indeed, many of the very greatest of these have been so carried away by the tender passion as to transgress the conventions and laws on sexual matters. In cases of precocious sexuality a feeling for the beautiful makes itself apparent; while with those who slowly mature, the æsthetic feeling is similarly delayed. But does not the infant who holds out a rose to you and cries "pretty," have a feeling for beauty? And it is surely unaffected by sexuality. What may be in the mind of a child speaking thus is hard to make out, but the activity is probably largely mimetic merely, and the term "pretty" is probably used substantively rather than qualitatively; it is the name of thing rather than quality. We certainly cannot assert of a child that because it uses certain words it attaches to those words the proper meanings. This is evident from the fact that a child taught to say "pretty" will bring you any and every object and use the word, or if it learns to take merely a class of objects, as rose, it does this at dictation. The child is, however, obviously attracted by some objects rather than others, but it would be hasty to say that it perceives their beauty, when it is quite sufficient to regard them as conspicuous only, and striking. But we have to touch on sensing later; and we 297only add to the evidence of connection of feeling for beauty with sexual feeling, that with the old and with eunuchs the æsthetic sense is but slight or tends to vanish. Thus positively and negatively there seems to be evidence that feeling for beauty originates in connection with sexual passion, either that the object of the passion is always regarded as beautiful, or that a feeling for beauty excites the passion. A girl adorns herself to attract lovers, knowing that to admire beauty is the first step to love. This close connection is recognised in common consciousness in that "lovely" is synonymous with beautiful, thus a "lovely" landscape or picture is a beautiful one.

That there is a close association of sexual with æsthetic psychosis is then obvious in the case of the human being, but yet it would be quite hasty to conclude that a sweet note or a pure colour may not be æsthetically appreciated by children before they have the first stirring toward sexuality, but still it is very easy—as I have before noted in the case of the child who cries "pretty!"—to mistake the quality of their interest.

But when we come to interpret the psychoses of the lower animals in connection with sexuality we may still more easily slip into a doubtful automorphism. Thus to say with Darwin, "When we behold a male bird elaborately displaying ... before the female, ... it is impossible to doubt that she admires the beauty of her male partner" (*Descent of Man*, p. 92), or more strongly still with Grant Allen, "Every crow must think its own mate *beautiful*" (*Mind*, v. 448), we too easily take for granted that these birds would feel like ourselves in corresponding circumstances. We can find a more simple explanation. That crows often maltreat those who are off colour, *e.g.*, white, plainly does not require us to suppose that they regard white as ugly, black as beautiful, any more than we should judge that students in some Society who wear a 298black badge would be æsthetically moved when they look with disfavour upon students who may wear a white badge. Animals are clannish, and as a rule, suffer none but those who have the customary marks to associate with them, and especially to propagate. Hence when the peacock displays himself to his mate he simply shows to her that he has most conspicuously the proper marks, and she sees that he is the proper mate. These are signs of a tempting mate, just as here is tempting food, a very red ripe berry, but the coloration no more in the one case than the other awakens feeling for beauty. The hen bird probably appreciates a red feather as a red berry merely as being signs of the completely satisfying. Sexual selection, like nutriment selection, is a discrimination according to certain characters as prompted by appetite. The expanded and vari-coloured tail of a peacock is then a mere sexual characteristic which does not imply feeling for beauty in its appreciation as significant of sex. A small foot, long hair, and other sexual characters in woman, which are attractive to men, in like manner arouse emotion which is far from æsthetic. We may take a perfectly unsexual æsthetic pleasure in long raven tresses just as we do in an ebony table, but this is obviously rather late achievement.

In fact are not æsthetic and sexual feelings mutually exclusive? So far as nude art is "suggestive," so far is the feeling of its beauty lost, hence sculpture is not tinted. And so in the presence of the nude model the artist can have merely æsthetic emotion, whereas his visitor is apt to have emotions of another sort. We do, indeed, say that the lover dwells upon his mistress' "beauties," but beauties here mean attractions, and to the devoted lover all parts are attractive, even moles and freckles which to the æsthetic eye are ugly.

From the evidence in hand we judge then that it is certainly not necessary to call in the feeling of the beautiful 299as the motive in the origin and development of sexual characters in animals and plants. Just as there is a cry of fear or a tone of anger there is a vocal expression of sexual feeling and emotion which has its use and is recognised as such, but whose æsthetic quality is no more a matter of immediate apprehension than in other utilities. At least the safest interpretation that we can now make for all the lower grades of sexuality is that sex characters are not primarily determined by the feeling for beauty, but are simply immediate signs of sex to awaken the sexual response and secure the best mate. How is it that sexuality is so prominent in expression among some species and so little among others?—compare peacocks and blue jays—is a question on which we have no light. We are also in ignorance how the particular sexual character was evolved and not some other, for example, why is not the peacock's tail red? Grant Allen's suggestion that food selection has influenced sex selection may be true, but it would require a very wide and thorough investigation. Do brilliant-hued birds prefer brilliant-hued foods? How is the coloration of the scarlet tanager related to the coloration of its food? However, if the colouring of foods and mates were the same, it would in some cases lead to disadvantageous confusion, and on general principles we should expect such distinct elements as nutrition and sex to develop on very different lines. The cue for colour may be learned first with reference to food, but it may be carried on as sexually significant on very distinct lines. Still to distinguish a food or a mate by colour is equally non-æsthetic in itself. At least we think it improbable that æsthetic psychosis arises as incentive to or reflex of sexuality in any of the lower psychic stages.

A theory of the origin of æsthetic psychosis which has been pressed by some, as by Herbert Spencer, is that it arises as reflex from spontaneous outflow of energy, or more particularly in connection with play impulse. A 300horse turned loose in pasture may gambol, running, sniffing, looking around, all which denoting a free outflow of energy through lines of least resistance, the customary channels of activity. But we cannot seriously think that in this sensing and muscular activity there is implied any real æsthetic psychosis, and indeed it seems quite emotionless. The emotion of fear or similar feelings aroused the original activities, but this present galloping, etc., is automatic, and such immediate pleasure as may result from this free activity is scarcely of the æsthetic order. The whole is of a distinctly lower order than the original activity and much below æsthetic quality. If we recall our own state of mind in youthful "letting off steam" and in plays, we do not find æsthetic pleasure. There is, however, a pleasure of relief and also positively a pleasure from such spontaneous outflow; but the outburst of pent-up energy automatically spent along lines of race action is a mere echo, dies out at once, and as degenerate form is not a starting point for origin of any new psychosis. Play as simulation of feeling and action is also removed from æsthetic activity, as in a dog playing at fear and running, or at anger and chasing. He gets a more or less modified fear or anger, but there does not seem to be any tendency to æsthetic psychosis. Mere imitation is more or less exact and skilful, but emotion therein and thereat is plainly not the glow of æsthetic emotion, but is reflex of sense of power and intelligence as qualities. Mimicry as mere outlet of energy as with monkeys is plainly not aesthetic; here is merely an automatic outflow of force into suggested activity. When a savage as mimetic achievement carves the figure of man as handle to a knife, he accomplishes art, but not fine art. He has no more æsthetic feeling than a boy or man whittling out a ship, it being merely an exact and skilful counterfeit of a real thing. Imitation for the sake of imitation or to deceive is a teleologic pleasure distinct from æsthetic. Successful imitation is often said, indeed, to be 301"beautifully done," but this means no more than well done. Even a well-baked cake is popularly spoken of as beautifully done.

We observe that superfluous energy rushes out along customary or habitual lines of activity, and so with perfect ease and economy. Activity which is easy and free is in itself pleasant, and this pleasantness in sensing and derived psychosis is æsthetic feeling. Where sensing is mere escape valve of force, though facility is absolute, there is, as just pointed out, no æsthetic quality, the whole tending to the merely mechanical. Owing to the fact that in nature curved lines predominate and so ocular adjustment is to them, my eye follows a curved line easier than a straight one, hence when spontaneous energy outflows in sensing activity of least resistance it will be toward curves. But spontaneous activity of this kind is, as we have explained, not æsthetic. The law of economy in a vent is, greatest force, least effect, the contrary of the usual formula for economy which is, least force, greatest effect. Where energy is expensive the latter rule is to be applied. Thus in directed and effortful sensing activity economy means the ratio of efficiency, the ratio of the amount of painful effort to desired result. But this is merely a saving of pain and not a real pleasure psychosis. When I, in using a microscope see clearly with less and less effort the objects of my study, I may take pleasure in the economical and facile accomplishment, but this pleasure is one of satisfaction in power and skill, and so not at all æsthetic. Again, a dyer has great skill and easy appreciation with respect to colour, but the æsthetic side of colour is not thereby specially felt by him. Mere habitual and easy colour sensitiveness is not then thereby æsthetic. We must, indeed, sense a colour before we can feel its beauty, but the feeling of beauty is not directly involved in any stage of the sensing evolution from the earliest and most painful effort with bare appreciation to the spontaneous 302and effortless sensing at the moment of great surplus of sensing energy.

Another way of accounting for æsthetic psychosis is by association. Pleasant sights, for instance, are those with which we associate pleasure, and "pleasant" means to many, beautiful. But a traveller, thirsty in a desert land, declares that he saw no more pleasant sight than a mud hole, but this pleasure, as he himself would aver, was far from æsthetic. Whatever we have associated pleasure with, we regard with pleasure, but only as we have associated æsthetic pleasure with it do we regard it with æsthetic pleasure. Thus mere association or revival no more gives us the derivation of æsthetic than any other emotion. Any pleasure or pain may be associated with any sensation or perception, and thereby re-occur with these, but the mere revival obviously does not alter the nature of the psychosis or give any new psychosis. It is not what is recalled, but how we *feel about* it that constitutes æsthetic emotion. So also when the beautiful is defined by H. R. Marshall as "the permanently pleasurable in revival," we get no insight into the origin, nature, and development of the æsthetic psychosis; this purely objective description gives no psychological analysis. But we may question the accuracy of the description. A thing of beauty is not a joy for ever when we mean thereby the object which excites the æsthetic psychosis, for much that has seemed beautiful to one people and age does not remain so for all peoples and times, and even with the individual, taste varies. We must also note that the permanently pleasurable in revival may not be æsthetic, as the lover's remembrance of a trysting place. On the whole, I do not find that æsthetic pleasure is in any case to be ascribed to association, though it comes under the general laws of association like any other feeling. A lily excites various modes of æsthetic impression by its form, colour, odour, poetical character, etc., all which may re-awaken together upon any presentation or suggestion 303of the lily. However, for the aboriginal lotus-eater the lily was also a pleasant sight—but not æsthetic—from the associated pleasures of its pleasant taste and as satisfying hunger.

We have implied throughout—and common introspection approves this—that æsthetic pleasure and emotion is a distinct psychosis which somehow arises with reference to objects. It is not some previous psychosis as modified by association, habit, economy, play-impulse, or sexuality; but it is a *sui generis* mode which develops on the basis of a past evolution. The simplest and earliest æsthetic mode is plainly the sensuous. Very commonly when looking on the delicate solid-tinted glow of early dawn I have æsthetic pleasure, my eye dwells on it with pleasure and drinks in the pleasant light. It is obvious here that the sensing activity is carried on, not to discriminate food or mate nor yet as mere vent to energy; but the sensing here acts for the pleasure in the activity itself. How and why mere cognitive act, which originates as guide to life, acquires a direct pleasure value and so is carried on apart from the ends of life, and initiates an æsthetic world of its own, cannot on the face of it be explained by natural selection; it is entirely apart from this order of things. But we know that sensing often carries pleasure with it as significant of life value, thus the thing tasting good was originally the good thing to eat, digest and assimilate; so also for smell, etc. But under natural selection this pleasure sanction and index was never cultivated for its own sake.

Now is there any real difference in the pleasure in, for instance, smelling, for the pure pleasure of smelling, as a perfume of fresh apples, and the pleasure from smelling the apples as detecting them when you are hungry? "How pleasant those apples smell! I do not care to eat them, but I just enjoy smelling them"; is the pleasure thus indicated the same in quality with that of the man who says, "Those apples smell so nice I would like to 304try one"? Again, if hungry, we say, "The bread tastes so *good*," but we notice this pleasantness rapidly decreases as appetite is satisfied. However, if there be fresh grass butter, you may continue to eat long after appetite is satisfied, for the pure pleasure of the taste. Obviously, the latter pleasure is not a mere continuance of the former. Relish and taste pleasure seem distinct. Again, a red apple is a pleasant sight to a hungry man and to an artist in different ways. If our pleasure in looking at a picture of an apple is such that the mouth waters, we know at once that the pleasure is unæsthetic. He who is very fond of apples, and to whom they are always a pleasant sight, is so far barred from æsthetic pleasure in them; while he who has no appreciation of their edibility is thereby prepared for æsthetically sensing them. So also sour grapes are as pretty as sweet. The colour sense began as discriminative of foods, and hence red became pleasurably known, but æsthetic appreciation is certainly much later and quite diverse. If it be asked how and when did red, already noticeable, become dwelt upon æsthetically, all we can hazard in reply is that at some leisure moment when unmoved by appetite a surplus of energy set up an habitual sensing activity, as noticing reds, and at a certain stage when some directing is exercised, there comes a unique pleasure from the mere sensing, and the red is therefore *dwelt* upon. Æsthetic colour-pleasure in the simplest case arises then in every one's experience.

Sense-pleasure is thus distinctly of two kinds, first, as arising in direct connection with general organic demands and satisfactions—the part as serving the whole; second, as arising immediately from the sense-activity—the whole as serving the part. A monkey may find an apple a pleasant sight, but loses all interest when the apple is seen to be an imitation: the monkey has the first pleasure, but not the second. The sensuous æsthetic problem is merely to introspect the quality of the sensing-for-itself-pleasure 305as distinct from pleasantness coming from the service of life. A sense which develops its own pleasurableness is on a new line, which we term the æsthetic. Æsthetic activity is distinct from mere vent activity of superfluous energy by reason of being carried on self-directed by the felt pleasure of activity; it implies a measure of self-direction and self-consciousness. Æsthetic activity may then be generally described as primarily a sensing carried on, not as means, but for its own sake in pleasure immediately resulting. And we find that in this very general meaning all senses have their æsthetic activity. The temperature sense is carried on, as in basking, for the pure pleasure of warmth. A cat behind a stove is a connoisseur in æsthetic warmth sensations, and enjoys warmth for its own sake, so far as often to injure the organism as a whole. To lie in the sun and experience the thrills of pleasurable warmth and to keep up this sensing merely for the sensation pleasure is a frequent experience even with man. Again, the muscular and pressure senses often have a sphere of æsthetic activity with athletes and lovers of exercise. When in prime condition, a man will toss weights about solely for the pleasure involved in the sense of pressure and of muscular activity. Touch also is plainly æsthetic when one handles silk for the pleasure involved in its smoothness. Smell is obviously an æsthetic activity in smelling perfumes for the pleasure of the smell. It is probable that the æsthetic activity of this sense is far wider in some of the lower animals where the sense is much more acute, as the dog. The dog is plainly having a very different psychosis when he is smelling with pleasure a piece of meat which he is about to eat, and when he sniffs carrion and perfumes himself therewith. He gets thus a certain pleasant but gross stimulation quite akin to the pleasure some men take in musk, an enjoyment of which is distinctly an animal trait. Again, the epicure who sips his rare wine is tasting for the pure pleasure of 306the taste, and exercises this sense æsthetically. The æsthetic of all these senses may be called the lower æsthetic, in contradistinction to the higher æsthetic of sight and hearing; but æsthetic activity is throughout its whole range practically identical in nature and in the quality of its pleasure. When I lie in the sun and get warmth, not because I am cold, but for the mere pleasure of the warmth thrills, and when I keep looking at a delicate tint in the evening sky for the mere pleasure of the sensation, I have, as far as my introspection assures me, activities whose method and pleasure tone is identical.

Simple sensuous æsthetic is no doubt the beginning of æsthetic activity, but there speedily enters much complication. It often happens that single elements which separately do not excite us æsthetically will produce a marked effect in conjunction, as complementary colours, for instance. Indeed, relation plays so large a place in our æsthetic experience that such principles as variety and contrast, or, on the other hand, unity, order, proportion, and harmony, have been made fundamental to the æsthetic feeling. Æsthetic effect certainly here becomes a complex of two or more reinforcing sensations or perceptions. Where the sensuous elements of a perception are in themselves pleasing we may expect the unison in perception to be doubly pleasing. However, we may also conceive that æsthetic pleasure arises as a reflex of perceptive activity in and for itself as a co-ordinating of impressions.

Fechner has made some experiments on what combinations are pleasing; but experiment in this direction is extremely difficult because so few people are willing to speak frankly of their æsthetic feelings, being very sensitive about compromising themselves on matters of taste. There is also the great difficulty of isolation, of making sure that association does not creep in and add unforeseen elements. If Fechner expected to get any judgments of value on such a matter as the golden section rectangle, he 307should have consulted only trained artists who are used to taking up the æsthetic activity with reference to any material and expressing themselves with freedom. If this rectangle has the æsthetic quality Fechner's experiments suggest, it seems strange it was not adopted by the symmetry-loving Greeks in their temples, like the Parthenon.

To the spheres of simple and relational sense beauty we have to add a third—representative beauty. A colour, or two or more in combination which give æsthetic satisfaction, will also please in hallucinatory vision and in representation proper where the revival is recognised in its unreality and representative nature, and also in recollection where the memory is willed. The mere imaging these colours without any definite time relation also gives æsthetic pleasure. It is, indeed, a pleonasm to say that æsthetic revivals are æsthetic. However, imagination is productive as well as reproductive, hence the ideal achieves a fuller beauty than the real. Where the mind, prompted by æsthetic desire, determines its own object, this object can more fully satisfy it than reality, which is always imperfect. Thus art surpasses nature, or more strictly is a higher nature. Idealism then is a mode of realism, and realism is but the ideal of actuality. But the imaging activity may, like the perceptive, be considered as in itself a source of æsthetic pleasure. Imaging is primarily used in the service of life, as when walking in a forest I hear a peculiar cry, imagine a wolf, and flee. When imaging has been largely developed thus, it may often act as a mere vent to energy; but this kind of activity has here, no more than elsewhere, real æsthetic quality. At the animistic stage children imagine in this way long before they æsthetically image. When we consciously and with some self-direction enjoy imaging for its own sake, we attain the æsthetic sphere. The æsthetic pleasures which are suggested by such a phrase as—

308"Fair ship, that from the Italian shore
Sails the placid ocean plains"—

are not merely the sum of the original sense pleasures, but perceptive and imaginative pleasure *per se* is added, the image is more beautiful than the real vision, and this perception than some sense element, as the light sensation implied in "placid."

Æsthetic pleasure, even in sense, and much more in perceiving and imagining, is a *delight*, that is, æsthetic quality is an emotion quality, it is not a mere feeling from an object, but a feeling about it. Now emotion may be enacted for emotion's sake and so an æsthetic pleasure wave be generated. This is the pleasure we take in the pathetic—pity, the sublime, fear as awe, the tragic-horror. These emotions are realized for themselves as a mode of pleasurable activity. Æsthetic emotion is also very largely emotion at emotion, as a feeling for the expressive, still here the emotion is for its own sake.

Æsthetic activity may then be described as an independent self-activity of some sense, or of perception, or imagination, or emotion as impelled by a pleasure, this pleasure being a distinct and new form we term æsthetic. It is probable this pleasure first arose in connection with the exercise of the sense as a vent for spontaneous energy, and pleasure once somehow being taken in a mere activity *per se*, it is thenceforth conducted therefor. This is the plainest path of conjecture thus far. If the first æsthetic pleasure were taken in some quiet moment of venting energy in sensing red, then red will continue to be sensed, impelled by the pleasure involved in the act. Granted such an origin, the development of æsthetic psychosis can be traced in the way we have noted.

Æsthetic psychosis is commonly regarded as passive, and it is indeed true that the first moment of the pleasure *comes* as result of an activity impelled by other motives. New psychoses are not consciously formed but are rather 309hit upon in natural development; but once a new pleasure is felt its conditions will be attained and kept to by conscious effort, and the pleasure itself will receive its development only through effortful activity. It is by supreme effort the great artist attains the vision of beauty, it is by supreme effort he expresses this vision, it is by supreme effort the critic appreciates this expression. He who has no appreciation of sculpture may by patiently and earnestly observing statuary reach at length some æsthetic pleasure. Thus the æsthetic, like all mental modes, so far as progressive, is effortful; and it seems certain that the æsthetic pleasures that come to us so easily are race acquirements, a heritage of culture. From its first germ onwards æsthetic, like intellectual, like moral, like all mental activity, is the achievement of intense struggle.

With the rise of beauty we have a new utility. Here is a new pleasure which once experienced is sought and sought again, is developed, and with some natures becomes absorbing passion, the life. Objects fitted to give this pleasure are desired, are bought and sold. The beautiful is used to effect all kinds of ends. The lover adorns himself to make himself attractive, the advertiser distributes his bills in artistic shape, the real estate dealer ornaments his houses and grounds. Whatever will afford æsthetic pleasure we are willing to pay for and pay high. In fact, in the person of a Patti the æsthetic thrill becomes the most expensive taste which humanity can indulge. Art then is a utility—a something which satisfies desire— and as such it is not free or shareable. But one at a time can observe a picture from the best point of view. Rich men buy the most sightly spots in nature, the places of magnificent vistas and open to beautiful sunsets. Beautiful things are then desirables just like edible things or warm things, and as such they are not shareable. The feeling for beauty, just because it is self-contained, is far from being disinterested. It is essentially selfish.

CHAPTER XVIII

THE PSYCHOLOGY OF LITERARY STYLE

Mr. Herbert Spencer's famous essay, entitled, "The Philosophy of Style"—by which is meant the Psychology of Style—propounds what we may term the economic theory of literary effect. The secret, he tells us, of the pleasing effect of diction, rhythm, figurative language, sentence structure, lies in this, that these are labour-saving devices to economize mental effort, that by their use we get with the least attention the greatest apprehension; and hence we receive pleasure as reflex of the facile and full cognition functioning. Literary pleasure is thus brought under the law of pleasure in general. Take the quotation from Shelley cited by Mr. Spencer:—

"Methought among the lawns together
We wandered, underneath the young grey dawn,
And multitudes of dense white fleecy clouds
Were wandering in thick flocks along the mountains,
Shepherded by the slow unwilling wind."

You have read this with pleasure, and is not the source of this pleasure the ease and celerity with which the mind reaches the "desired conception"? Vividly and forcibly the mind is led by cunning use of phrase and rhythm and figure to realize the picture, and there is a glow of pleasure in the reaction from the facility. Language is a medium for the transfer of ideas, and when it accomplishes this office most effectively, as in the present case, and acts upon the mind so clearly and forcibly that *nolens volens* 311the reader at once apprehends and comprehends, he feels a thrill of pleasure therewith, just as there is pleasure connected with the rapid and easy assimilation of well cooked food. Before developing and criticising this theory I may remark in passing that Blair, the rhetorician, in treating of the structure of sentences foreshadows in a way the economic theory when he writes that "to have the relation of every word and member of a sentence marked in the most proper and distinct manner, gives, not clearness only, but grace and beauty to a sentence, making the mind pass smoothly and agreeably along the parts of it." This surely implies that æsthetical pleasure of style may be based in a psychological economy and facility. It is indeed a commonplace remark, "The book is so well written that you cannot mistake or miss its meaning"; wherein the identification of style with intelligibility becomes a truism. Certainly Mr. Spencer has not in the economic theory propounded anything radically new.

We note at the outset that while this pleasure of style may result from economy it is not the pleasure of the conscious economizer. The reader who is enjoying a very readable book has a distinct pleasure from him who views with satisfaction his finishing a book at a great and unexpected saving of mental energy. We have here the direct pleasure from economical exercise of the faculties contrasted with the indirect introspective-retrospective pleasure at economy effected. Many persons take as much pleasure in making mental energy go as far as possible, but this pleasure in economy is obviously not the pleasure of style, which is not reflective, but naïve and direct impression.

Language, either spoken or written, by its more or less effective modes of accomplishing its office does then awaken a simple and direct pleasure, according to the general law that pleasure accompanies efficient acts as a sanction and stimulus. It is obvious that style for spoken 312language, oratorical style, is precedent in its formation to style for written language or literary style, and that it has greatly affected literary style throughout its whole history. Yet the distinctness of the two modes is affirmed by the common observation that a speech, impressively pleasing to listen to, often does not read well. While it may be true that in its origin literary style borrowed certain devices from oratorical, yet in its latest evolution the written page is far from being the speaking page. The book is not a substitute speaker addressing us, and modes of expression which are most fitting for conversation and oration, though sometimes used by writers, are alien to pure literary art. However, I cannot pursue this interesting subject, nor yet can I here treat of the origin of style more than to merely observe that it is considerably later than the origin of language itself. Neither the original uncouth speech, whether interjectional or onomatopoetic, nor the earliest rude inscriptions can be said to have style, oratorical or literary. Style is the offspring of specialization; it first appeared when men recognised some one as particularly gifted for fitting expression, and chose him as spokesman because of this ability to communicate what was desired to be said with special force and clearness. Thus arises the orator who achieves and invents oratorical style. Likewise the writer is one who is selected for his special abilities in expression by word of pen, and the scribe, clerk, and public letter writer arise and evolve literary style as a skilful way of effectively conveying ideas and impressions by written language. The reader is also evolved, and in the reciprocal relation of demand and supply and the competitive struggle to secure readers, the writer seeks ever more and more to please and interest by introducing and perfecting various inventions to make the reading of his work very easy and enjoyable. Thus it comes that readableness is the natural test for reading matter.

The economic theory of style in fine art plainly implies 313at bottom physiological economy, for all psychological economy can only be effected on this basis. The psychology of style must rest on a physiology of style. We know that the pleasures of form and colour in sculpture and painting are the reflex of physiological functions as easily and completely performed. The curve of beauty is such because the eye follows it more easily than other lines; the pleasing colour is such because the physiological stimulus is accomplished in a normal and facile way. And as visibility is the test for the arts which appeal to the eye, so audibility is for the fine art which appeals to the ear. Pleasure from music is the reflex of aural functioning accomplishing the most with least strain. Now the pleasure which comes from literary style must similarly be sought in some physiological mode. While plain print and good paper are incidental pleasures in reading, they are not primarily due to the stylist, who does, however, appeal to the eye by the due proportioning of long and short words, sentences and paragraphs. Though there is no conscious intent by the stylist, yet it may be believed that the use of certain letters and certain successions of letters as more or less easy for the eye is a matter of some importance. Some letters and some combinations are ocularly more pleasing than others, and this is clearly founded on economic physiological conditions. It is greatly to be desired that physiologists would invent new alphabetical forms which should be most adapted to the eye. It is scarcely to be supposed that our present A B C's are the simplest and easiest line-combinations for the eye. When the visual side of reading is made as easy as possible, the general reflex sense of facility and pleasure therewith is certainly increased. The artificial languages now being exploited, as Volapuk, ought and would effect a great physiological saving, as would also be accomplished by a phonetic spelling.

But the direct visible function of style is certainly far 314inferior to the indirect. The power of style is very largely in stimulating pleasing visual images. The main element in literature we are told is vision and imagination, which is but a restimulation and recombination of ocular experiences. Sensation is the source and strong basis for all those faint revivals which are so aptly and pleasantly called up by the literary artist, and hence when the poet speaks of "the light which never was on sea or land," this is really meaningless, since all our light impressions are terrestrial in their nature. To the blind man the whole visual effect, direct and indirect, of style is lost; his imaging power must be in some other sense.

Literature is then, like sculpture and painting, largely a visual art, and its pleasure-giving quality is the reflex of visibility. Mere form and colour may in a sense constitute a picture; though in general we demand that it mean something, suggest something. A picture is such as depicting something, and so being more than a study in form or colour. The mere direct pleasure of ocular sensation plays a large part in graphic and glyptic art, yet it is commonly conceived that some measure of imagination, that is, some indirect visible function, is necessary even here. Sculpture and painting depend like literature on both direct and indirect vision as physiological and psychological basis of æsthetic pleasure.

But in a secondary way literary style depends for its effect upon auditory sensations both direct and revival. We mentally, and often orally, pronounce as we read, and so appreciate sonorous quality and onomatopoetic force. Alliteration, rhyme, euphony, and rhythm play certainly a considerable part in the charm of style, and literature on this side approaches and passes gradually into music. Euphony answers to melody, and rhyme and rhythm to harmony. Literature may become for us merely a succession of pleasing sounds, as when we hum over some favourite lines of poetry, or when, ignorant of the Italian 315language, we listen to an opera. Some of Milton's lists of names in such lines as these,—

"Of Cambalu, seat of Cathayan Can,
And Samarchand by Oxus, Temer's throne"—

charm merely by the flow and fulness of sound. But the stylist aims, not merely at formal sensuous beauty in tone and cadence of language, he aims to suggest pleasing sounds, and to awaken the auditory imagination, and to harmonize sense with sound as is done so successfully by poets like Tennyson and prosaists like Sir Thomas Browne. All this auditory side of literary style is lost on the deaf, as the visual is lost on the blind. Literature as an art is neither blind like music nor deaf like painting, but it is a compound art, visual-auditory, and thus, by virtue of its range, is the greatest of the arts. It is true that indirectly and in a very limited way painting can suggest sounds, and music sights, but literature, both directly and indirectly, can freely and fully give both. Word-music and word-painting are both methods of literary style. In short, the explanation of the pleasure of style is pleasing sight or sound directly or indirectly given, and the explanation of the pleasing character of the sight or sound is as the reflex of easy economical physiological functioning as basis of easy economical psychic function.

But we have now to ask whether economy of attention is the sole psychological secret of style, and whether, indeed, it is always necessary to style. Is style, like grammar or orthography, merely a more or less conventionalized device to make intelligibility certain and easy? Is our reading always the more pleasurable as it is the more effortless? The pleasure of facility certainly bears a large part in much of our literary enjoyment, but there is another and opposite law of pleasure which, I think, often determines pleasure in style. To accomplish much with no exertion, to slide down a long hill, gives pleasure, but there is also a pleasure in exertion, in climbing hills 316as well as sliding down. The pleasures of strenuous activity of attention form a certain element in literary effect. The writer may do too much for the reader, may make everything so simple and easy that the reader has nothing to do, but is carried along without volition and curiosity, losing all joy of attainment and grasp. For my own part, I often find authors too fluent and facile, especially among the French, and sometimes among the English, as, for instance, in some of John Stuart Mill's writings. These do not leave enough for me to do, and led skilfully along so smooth a road that I am not conscious of moving, I lose the pleasure of achievement, of the sense of enlargement of conscious powers. Easy got, easy goes, is the law here as elsewhere. The pleasure of acquirement is directly as the amount of attention exercised.

Mr. Spencer in discussing this matter remarks that, as "language is the vehicle of thought, we may say that in all cases the friction and inertia of the vehicle deduct from its efficiency, and that in composition the chief thing to be done is, to reduce the friction and inertia to the smallest amounts." But it must be remembered that motion is not only against friction but by friction. The rail may be too smooth as well as too rough. Every locomotive, for a given piece of track with a given gradient, has a certain co-efficient of friction for its most effective working, above and below which there is alike decrease of efficiency; and in engineering it is equally a problem to keep friction up as to reduce it. So I say of style, that it may be too smooth and facile, and may reduce mental friction to so low a point that there is no grasp and no real progress. A sentence of Hooker or Milton, magnificent stylists though they are, can, as an affair of economy of attention, be greatly improved by breaking it up into a number of simple plain sentences after the primer fashion, The cat mews, The dog barks, etc.; but this process certainly is not an improvement of their style. But if 317economy of attention were the sole secret of style, certainly the more economy we introduce the greater and better should be the style. Professor Sherman, of the University of Nebraska, in a recent article shows that heaviness—that which requires "constant effort in reading"—is due to the number of words *per* sentence, which has been reduced in the course of the history of English prose from an average of fifty words a sentence in Chaucer and Spenser to five in the columns of a modern, low-grade, popular story-paper; but it obviously cannot be maintained that the style of the story-paper is ten times better than that of Spenser's *State of Ireland.*

We might then set up with plausibility an exactly opposite theory to the economic, and maintain that the secret of style is in exciting us to the greatest attentive effort, and that the best style is that which rouses us to the severest mental exertion. However, I believe that these two opposite methods of style are complementary. The great stylist is he who strikes the exact mean between over facility and over difficulty, and touches the exact co-efficient of mental friction in the reader, at which his whole power of mind comes into highest and most harmonious and effective exercise. The accomplished stylist most cleverly throws in questions, suggests doubts, and defers answers. To read his book is not a toboggan slide, but an obstacle race. What is plot interest but a skilful putting of obstacles in the reader's way, deferring and thwarting his expectations, putting him on the *qui vive* of attention? By the development of plot the novelist and dramatist plays hide and seek with the reader. No cunning artist reveals at once his whole thought in a blaze of light, but he mystifies and draws in half-tones, thus to stir you to reach out and grasp his meaning.

But we are as yet far from exhausting the psychological significance of pleasure in style when we trace it to a reflex from either decrease or increase of attentive effort. 318The pleasure we have so far considered is naïve and direct; it is from literary art rather than in or at literary art as such. The child and the most ordinary reader derive from books a simple and natural pleasure which they do not reflect upon, and do not in any wise conceive the ways and means by which the effect is produced. Indeed, in the presence of the most lucid and perfect art these readers, like Partridge at the play, take everything as a matter of course, as just the way they would themselves express it. The *dilettante* alone tastes the pleasure in style as such; as an art, an adaptation of means to ends, he alone appreciates the delicate adjustment of expression to thought, the choice diction, the deft management of word and phrase. The quality of this technical pleasure in style is exemplified in its highest form in this note of a great artist-critic, Shelley, appended to his fine translation of the opening chorus in "Faust":—

"Such is a literal translation of this astonishing chorus; it is impossible to represent in another language the melody of the versification; even the volatile strength and delicacy of the ideas escape in the crucible of translation, and its reader is surprised to find a *caput mortuum.*"

The psychological nature of this pleasure in style is obviously quite distinct from the direct pleasures from reading which have been previously discussed. Here is pleasure in literary art, not for what it brings, but for its own sake. The distinction between the pleasure the average tourist takes in travelling swiftly and smoothly in a *de luxe* train, and that taken by the professional engineer inspecting the high-speed locomotive, is analogous in quantity and quality to the distinctive pleasures of critical and uncritical appreciation of fine art. But we have as yet only cleared the ground toward ascertaining the psychological *rationale* of literary style. We have marked only general causes of literary pleasure, we have noticed in this pleasure only those elements which flow 319from the psychological and physiological basis of all pleasure as reflex of functioning. That we admire and take pleasure in nice adjustment of means to ends is also a general law of pleasure with all who act teleologically, and are capable of appreciating actions of this kind. But is there not a specific quality in the æsthetic pleasure from or in literary art which has not yet been accounted for? Certainly the common expression, "more forcible than elegant," as applied to spoken or written language, denotes that for the popular consciousness style is somewhat more than and different from mere force and consequent ease and largeness of apprehension. We hear a very loud sound with greater ease than smaller sounds, there is economy of attention, yet this does not bestow æsthetic quality on the great sound. At the renderings of the finest music we are often called on to strain the ear, and the mental receptiveness as a whole to the utmost, in order to hear, note, and appreciate the delicate effects. So in literary art it is not that which speaks most loudly and strongly to the mind that thereby becomes the best style. In fact, the most forcible method of expression is often, as is generally acknowledged, slang, which is debarred from style. Literary style seems, then, more than a mental labour-saving machine. As a utilitarian device it certainly does save mental exertion, and gives rapidity, accuracy, and facility to psychic function. Like grammar, a mechanic rhetoric is useful, and we receive a pleasure from its use as from any other mechanism of man's industry; and further, we may take a certain pride and pleasure in its consciously recognised effectiveness. However, we have not yet reached style in the higher sense, which may be clear and forcible, but must be dignified, graceful, and beautiful. For purposes of business, for conventional communication, for science, for philosophy, language fulfils its end in stating accurately, clearly, and forcibly; but style as literary art is more than instrument 320to intelligibility, it has an independent office of its own. Language in the lower service as a medium of communication is a lens which cannot be too transparent; but in the higher service to fine art, language is rather a mosaic window of stained glass which both absorbs and transmits light, which both conceals and reveals, which we look at as well as through. In literary art or style, language has a value of beauty for itself alone, as well as a value of use as a means of communication.

But the root of style is in emotion; it is as expression of emotion, and in the main of one kind of emotion, that language rises to style. All emotions influence language expression, and any one may, under certain conditions, lead towards literary art; there is an eloquence of wrath and of fear, of hate and of love, and these emotions may induce artistic creativeness in written language; but the main impulse to art is in the feeling for beauty *per se*. This is a certain mode of emotional delight which every one who has felt it knows at once in its quality as quite distinct as a psychic mode. How literary style rises and falls with æsthetic emotion might be exemplified by a wide range of quotations, but an example or two must suffice. This, from one of Shelley's letters, will, I trust, illustrate the point:—

"My dear P——, I wrote to you the day before our departure from Naples. We came by slow journeys, with our own horses, to Rome, resting one day at Mola di Gaeta, at the inn called Villa di Cicerone—from being built on the ruins of his villa, whose immense substructions overhang the sea, and are scattered among the orange groves. Nothing can be lovelier than the scene from the terraces of the inn. On one side precipitous mountains whose bases slope into an inclined plane of olive and orange copses, the latter forming, as it were, an emerald sky of leaves, starred with innumerable globes of 321their ripening fruit, whose rich splendour contrasted with the deep green foliage; on the other the sea, bounded on one side by the antique town of Gaeta, and the other by what appears to be an island, the promontory of Circe. From Gaeta to Terracina the whole scenery is of the most sublime character. At Terracina precipitous conical crags of immense height shoot into the sky and overhang the sea. At Albano we arrived again in sight of Rome. Arches after arches in unending lines stretching across the uninhabited wilderness, the blue defined line of the mountains seen between them, masses of nameless ruin standing like rocks out of the plain, and the plain itself, with its billowy and unequal surface, announced the neighbourhood of Rome. And what shall I say to you of Rome? If I speak of the inanimate ruins, the rude stones piled upon stones which are the sepulchres of the fame of those who once arrayed them with the beauty which has faded, will you believe me insensible to the vital, the almost breathing creations of genius yet subsisting in their perfection?"

This letter opens with language as method of conventional commonplace communication. The second and third sentences are barely tinged by æsthetic emotion, as in "immense substructions" and "lovelier"; but it is not till the fourth sentence that style fairly begins. Then it rapidly falls away in the fifth, sixth, and seventh sentences, to arise again with a new wave of æsthetic emotion, which progresses through the remainder of the quotation. The culminating points of the æsthetic emotion are precisely the culminating points of style, namely, in the phrases, "an emerald sky of leaves, starred with innumerable globes of their ripening fruit," and in "sepulchres of the fame of those who once arrayed them with the beauty which has faded." What constitutes the peculiar attractiveness of these expressions is this, that they are rich in æsthetic feeling, and communicate it to us. We are by 322the power of style sharers in high delights. In the first case we are awakened to a visualizing, to a sensuous beauty, though compounded with other elements, through metaphor; and in the second case the emotion is a complex of sensuous and spiritual elements.

Take also the verses from Shelley already quoted. Mr. Spencer, in commenting on these lines, has correctly pitched upon the word "shepherded" as the culminating point; but when he intimates that the beauty and pleasing effect is due to the "distinctness with which it calls up the feature of the scene, bringing the mind by a bound to the desired conception," we must dissent. This purely utilitarian explanation fails to recognise that poetic metaphor is confusing—here two classes of objects, clouds and sheep—and misleading, except to the poetic mind. A writer who was aiming purely at clearness and correctness of imaging, as a popular scientific writer, might mention the clouds as like patches of white wool; but he would not bring in the extraneous ideas of sheep and shepherd. If Mr. Spencer were trying to give us a vivid idea of clouds, he would surely not speak in this purely poetic fashion. It is a mode of fancy and emotion which the poet is indulging when he writes these lines, and not an intellectual impulse to clarify and illustrate. If Mr. Spencer receives them in this latter spirit, he misses their psychic content and explanation. Poetry is only intelligible to the poetic, and the German pedant who emended "Celia, drink to me only with thine eyes," to "Celia, wink to me only with thine eyes," was certainly economizing attention and rendering conception easy, but at the expense of poetic beauty. The source of the pleasure we take in poetic style—the highest and purest form of literary art—is evidently not for its intelligibility, at least primarily, but its æsthetic quality, an expression of a peculiar emotional attitude toward objects.

To illustrate this psychological distinction between the 323sense of beauty as inherent in style, and style as mere force and clearness, I instance further only this sentence from Mr. W. D. Howell's Italian sketches, describing a side wheel steamer in motion: "The wheel of the steamer was as usual chewing the sea, and finding it unpalatable, and making vain efforts at expectoration." This is the *ne plus ultra* of a *pseudo* literary style, of affected and strained literary art. An ugly metaphor, forcible and clear enough, is relentlessly pursued to its ugliest conclusion. Here is style in pin feathers, and we are glad to remember that it was writ in callow youth. It brings "the mind by a bound to the desired conception," but this does not sanction it as fine art, for it is utterly without taste and beauty.

I believe then from considering the previous examples—and they might be indefinitely extended—that the main function of literary art is not intelligibility, and that pleasure in style in its specific quality does not arise out of economy of attention, but it is a direct communication of pleasant æsthetic emotion artistically conveyed. Intelligibility is a regulative by-law of art, but it is neither standard nor goal. Literary art is then a compromise between intellectual and emotional motives, between sense and sensibility. The natural choice and order of words for easiest apprehension is rarely the artistic order, as every *littérateur* knows full well. It is, for example, simplest and clearest to repeat the best and exact word, yet the literary artist avoids, and rightly, the repetition of words in the same sentence or paragraph. Thus also, while, as Mr. Spencer suggests, rhythm and euphony may often help sense, yet I believe they as often distract from it. We often tend to turn over in a very senseless way words and verses which please the ear. As language is both an organ for meaning and for beauty, literary art, like architectural, is always a compromise between utility and beauty, that is, neither literature nor architecture are pure and perfectly independent arts. However, it is possible 324that poetic license may, as has already been done to some extent in English, ultimately develop a pure poetic language, entirely distinct from the utilitarian product, and bound by none of its practical rules; then and then only will literature become a pure art.

Further, that literary art does not always imply clearness and consequent economy of attention is evident when we reflect that the nature of emotion is to disturb the mind, and hence also the language expression. Incoherence, dimness, darkness, as qualities of æsthetic emotion, render literary art correspondingly broken and obscure. The weird, fantastic, and mysterious issues in style which is far from being easily intelligible. In the dreamy poetry of the Orient all is hazy and evanescent, and the mind strives in vain for clear impressions, yet here is the peculiar charm of style. Among Occidentals William Blake, with his childish incoherence, and Robert Browning, with his harsh abruptness, have a certain obscurity, but both are great stylists and great poets.

Style then is at bottom something quite distinct from either ease or difficulty of apprehension. It is founded, not on apprehension at all, but on emotional receptiveness. Hence very active and intellectual natures seem ever debarred from really entering the realms of art, because they ever fail to appreciate that the function of art is not practical, or ethical, or scientific, or philosophic, but emotional. The man of business, of politics, of science, of thought, cannot give himself up without questioning to be thrilled and suffused by the unanalyzable charm of mere beauty. Such natures seem incapable of receiving, they must get and acquire, and so they miss all that art to which the only open sesame is a quiet inattention and a wise passiveness. The kingdom of art is not taken by violence, and the violent do not take it by mere intellectual force.

As to the origin and nature of the feeling for beauty 325in style as for beauty in general, the reason may be sought in survivals of primitive pleasures. Thus the expression, before quoted, "starred with innumerable globes of their ripening fruit," aside from the pleasure in sonorous quality and artistic construction, pleases mainly as awakening the feeling for natural beauty. But what is the psychological explanation for this æsthetic emotion in presence of tree, fruit, flower, sky, and all landscape features. It may largely be a revival of feelings felt long since by our arboreal and forest-haunting ancestors, "combinations of states which were organized in the race, during barbarous times, when its pleasurable activities were chiefly among the woods and waters" (Spencer, *Psychology*, Sect. 214). In the woods and by the streams there tends to revive the long outgrown physical emotion; the old savage feelings of delight and excitement in the chase come back to the civilized man, and in stealthy approach of game and skilful slaying the modern man re-experiences far distant ancestral joys. Now literary art by skilfully setting forth scenes of savage life may renew, the old survival feelings to a certain degree of illusive life. This is done to a large extent by pastoral poetry, mythic story, legend and fairy tale, whereby we drop back into a very old and simple mode of enjoyable mental life. The basis of primitive psychosis is in the particular concrete and animate, and literary art, especially in its highest manifestation, poetry, as becoming simple, sensuous, and impassioned, has a foundation in survival tendencies. Through literature mankind renews its youth. Similarly we may suppose that if in the future psychic evolution of the race the present mode of thinking in general and abstract terms should be succeeded by some new and higher phase, then the artificial stimulating the revival of this outgrown abstract phase would constitute a source of pleasure and might be achieved through a style. As a means toward revivals literary style is a backward moving spirit in sharp 326contrast to science, which, as generalizing and depersonifying, is the forward moving process.

However, we have sharply to distinguish between what is given in a survival state and that which accompanies it. Primitive realization is always single and naïve, but when it comes up in a survival it is generally consciously contrasted with accustomed modes by consciousness, and there arises a reflective pleasure of contrast which is not contained in the survival itself, but of which the survival is merely a condition. Further, our realization of the outgrown psychic elements is very generally dramatic. We take self-conscious pleasure in investigating, assuming, and re-enacting past psychic phases. Even when a survival state arises spontaneously and naturally, it holds consciousness at best in its original *status* for a moment only, for self-consciousness quickly occurs and brings in a variety of secondary emotions. However attained, the obsolescent type of consciousness does not stand in its simple original force, but most often there is more or less make-believe, some sense of its artificial and unreal nature: we do not become children by playing at being children. Children and savages are in the animistic psychic stage, but the poetic interpretation of nature by adult man is plainly far more than mere revival of this stage, it is dramatic self-conscious realization. Original animism is often painful; the savage fears his gods and the child dreads ghosts; but myths and ghost stories are sources of amusement to us, and the twinge of fear which comes up as survival loses its real force and is dramatically realized and enjoyed. Literary art is a dramatic induction into the past rather than incentive to mere revival; and it makes us to pleasurably renew alike the outgrown pains and pleasures. We certainly should go far astray if we should consider style as effectual mainly by its exciting to revival of ancestral experiences. What is recurrent is but a small element compared to what is concurrent.

327We must note the particular case of landscape beauty. Shelley's description of the orange tree laden with fruit excites in us the feeling of pleasure in the beauty of nature, a feeling which is declared by some to be merely the reminiscent revived feelings which our distant progenitors felt in the presence of natural forms and forces. But what was the emotion our remote progenitor felt at sight of a well-fruited orange tree? Did he feel moved as Shelley was and as we through Shelley are? and is our emotion but a faint survival of that which welled up in him at viewing the mass of green and gold, or has it any relation thereto? The civilized traveller in wild regions is often charmed by the beauty of the scenery which the savage natives do not in the least appreciate. But the revival feelings which come over him must be identical with the feelings of his unæsthetic companions who are totally insensible to natural beauty. The reversal tendency can give to the traveller only an animal pleasure in viewing an orange tree as satisfying to the taste and stomach; a fine, bright day can only suggest the pleasure of a sluggish basking. Goethe rejoiced that, though the incidental pains of æsthetic sensitivity were great, yet he could see in a tree shedding its leaves more than the approach of winter. Bare revival then cannot in itself constitute æsthetic pleasure or explain it. A savage race transferred to a civilized land for a few generations and then returned to their native haunts have acute pleasures of revival, but these are not of the æsthetic quality. An outcropping survival tendency may serve as itself an object for emotion and æsthetic emotion to the mind experiencing it, but thereby the survival is like any other object, physical or psychical, which excites æsthetic sensibility, and it no more explains the emotion for beauty than any other object.

It is evident thus far that the psychological basis of stylistic effect is very complex, and in this essay we 328certainly lay no claim to making an exhaustive enumeration of its factors. However, we have still to consider one more element, and perhaps, at least for cultivated minds, the most important psychic element of literary art. Read now the following extract, and analyze the impression it makes:—

"The natural thirst that ne'er is satisfied
Excepting with the water for whose grace
The woman of Samaria besought,
Put me in travail, and haste goaded me
Along the encumbered path behind my Leader,
And I was pitying that righteous vengeance;
And lo! in the same manner as Luke writeth
That Christ appeared to two upon the way
From the sepulchral cave already risen,

A shade appeared to us, and came behind us,
Down gazing on the prostrate multitude,
Nor were we 'ware of it, until it spake,
Saying, 'My brothers, may God give you peace.'"

Here, surely, is neither facility, nor beauty of expression, nor deft and subtle art to please the mind, yet it attracts and interests. The main secret of the effect of Dante's style is as revelation of personality. Art with Dante is the child of life, the product of long and deep-felt experience; and because he is an original reality he achieves in his writings that distinctiveness and distinction which is the truest and highest mark of style. Again, it is not the lucidity of Sam Weller's remarks that pleases us, but rather their characteristic flavour. We delight to come in contact with originals, and we relish the characteristic for its own sake, even when ugly or when most unlike ourselves in tendency, and so the modernest of the moderns enjoys Dante, the typical mediævalist. Style is the man. This is the best definition of style and the best explanation of its peculiar effect. Style is expression of subjective quality. While scientist and philosopher aim to be objective, to justly reflect and interpret outward reality the literary artist aims merely to give a perfect exposition 329of himself. Style is the literary expression of self-realization. Hence the greatest stylists write to please themselves, and are their own severest critics. Style is *timbre*, and the best style is that in which this peculiar tone of the individual mind is most perfectly revealed. A great style is, then, the expression of a great man, and the consummation of style occurs when the genius has grown to the highest point of his individuality—and individuality is genius—with corresponding power of expression. Among Tennyson's poems the most Tennysonian has the greatest style. When we quote from Wordsworth such lines as,—

"The world is too much with us: late and soon,
Getting and spending, we lay waste our powers"—

and say of them that they are eminently Wordsworthian, that no one else could have written them, we have said the highest word for the style.

In the very largest sense style is the evolution of the characteristic; development physical and psychical is but a movement toward style. The progress from homogeneity to heterogeneity in matter; the morphological development of animate things from indefinite formless beings to definite, complex types; biological integration and specialization—all this is progress of style. Thus the most lion-like lion and the most elephantine elephant respectively achieve the highest style of animal in their kind. The development in the human race is mainly psychic, and includes psychic classes, orders, genera and species, not as yet so clearly tabulated as in general natural history. A genius is the inauguration of a new *genus*, style, or type of man; he is a psychic "sport," to borrow a botanical term. A new mode of personality is achieved and may manifest itself in various ways of action, thought and emotion. If the expression is through literature a great style is generated, and this style grows with the growing individuality—the productions of youth have little style—and culminates with its culmination.

330To discover style is almost as rare a gift as to achieve it. The critical sense is about as uncommon as the creative power; hence the greatest masters of style have had often to wait long for recognition, which would hardly be the case if the main value of style was in economising attention. According to this theory, we should expect the stylist to be welcomed with instant and universal appreciation, a phenomenon which rarely or never occurs. With very many writers, as with Wordsworth, recognition is very tardy, and with some only posthumous. Many readers fail even with the utmost attention to appreciate the greatest artists, and can make nothing out of them; a few rise at length to some understanding; but only rare and select spirits find themselves at once *en rapport*. The true *connoisseur* and critic must introduce and interpret to us the characteristic quality or style of the *littérateur*, else we may never know and feel it. Recognition and appreciation of style as the characteristic is, then, for the vast majority an acquired taste; it is slowly and painfully learned, and so the emotion for style as specific mode of expression must be pronounced a very late psychic development.

The taste and emotion for the characteristic as such, whenever and however acquired, is certainly a peculiar and definite mode of emotion. It is far from being the feeling of discipleship, and is often excited by that which is most remote and opposite to ourselves. We say of a certain person, "He is a *character*," and he interests and pleases us as such, though entirely foreign to us in either sympathy or antipathy. As an entirely disinterested emotion, the æsthetic is beyond the range of common naïve consciousness. The enjoyment of the characteristic *per se* is specially for the analytically super-conscious cosmopolite and for the cultured critic. The pleasure comes partly from the novelty and the contrast reflectively understood, partly from admiration for the forcefulness of creative personality, 331 its plastic power in forming its material of expression, and largely a teleologic pleasure in perceiving fulness and purity of type. The emotion for style as characteristic expression is plainly one of those which is not due to the utility in the struggle for existence, but has arisen when experience comes to be cultivated for its own sake.

When, as in eras like our own, personality weakens, and the inner plastic and creative force of conviction and emotion decreases, the writer is driven to technical treatment. The *littérateur*, as he has little or nothing to say, contents himself with playing tricks on language, and elaborating rhythms and cadences. Style becomes finicky; a race of prinking poetasters and priggish prosaists arise, punctiliously formal, and superlatively dainty, who attain the art of saying nothing very elegantly, elaborately, and brilliantly. An over-conscious, over-subtle technique destroys the grand style as transmitter of characteristic quality.

I trust I have, in this brief study, made it clear that the psychology of literary style is far from simple, and that a number of factors are involved, which are slighted by Herbert Spencer and others of that school. I believe that any one at all conversant with literature who will reflect upon the pleasures he receives from reading, will perceive that the pleasure of smoothness and facility, of moving along rapidly and easily, is but one, and that generally a minor factor in literary enjoyment. Beside this, he often has the pleasure of difficulties overcome, of ideas grasped, and delicate emotional touches appreciated by triumphant attentive effort. Again, he receives pleasure in perceiving literary skill, the adaptation of artistic means to the artistic end. But, as I have maintained, the chief mode of pleasure is through style as transmitter of æsthetic emotion and as expression of the characteristic, achieving its acme when both these functions are simultaneously performed most fully and perfectly.

CHAPTER XIX

ETHICAL EMOTION

The need of a closer psychological definition and interpretation of ethical emotion must be apparent to any reader of the current psychology, where we find the utmost confusion and looseness of usage. One of the most glaring instances which I have come across is this from Perez (*First Three Years of Childhood*, p. 286): "As soon as the child begins to obey, from fear or from habit, he enters on the possession of the moral sense; as soon as he obeys in order to be rewarded or praised or to give pleasure, he has advanced further in this possession." A boy at table reaches out for the last piece of cake, but withdraws his hand out of love for his mother's approbation, and fear of her disapprobation. Does this imply moral sense and emotion? We say, indeed, that these were very proper and moral emotions for the child to have; objectively moral, but we do not describe the psychical state of the child correctly by saying that it has the moral sense and emotion. In fact, just so far as he acts out of love or fear, just so far he is not acting out of ethical emotion; that is, simply because he feels he ought.

Only the slightest introspection, then, is needed to recognise the distinction of objective and subjective morality, of a moral emotion and the emotion of morality. So we must disallow even dread of "moral discomfort" as psychically moral, Spencer notwithstanding (*Essays* I., p. 348). The fear of remorse may restrain from objectively 333immoral acts, but the ethical emotion is not a fear constraint, as every one knows when doing a thing simply because he feels he ought to. Because I judge my feeling or act a right one, does not constitute this the feeling of rightness as psychic fact. In short, we must always distinguish between the socially right action, the morally right action, and the psychologically moral action. He who erects a model tenement, even though he do it to advertise himself, is doing the right thing by society, though his action is neither prompted by a moral emotion nor the moral emotion. If philanthropy incites him, both the act and feeling objectively are moral, but psychically he is immoral, and only becomes psychically moral when he acts out of the ethical emotion as feeling of duty. One who acts out of sympathy, pity, mercy, affection, feeling of honour, love of approbation, and similar emotions, often confounded with the moral emotion, is objectively moral. We pronounce these to be right emotions, yet they are not the emotion of right, and so not psychically moral; and it is evident, also, that they may not be socially right, for often actions from these motives result in social wrongs. However, in later phases of psychic evolution, when emotions themselves are reflected upon as psychic acts, the emotion of the moral "ought" may be felt as stimulus to them, and so we may at once feel that we ought to sympathise, and so sympathise, and so act, and may thus at the same time be psychically, morally and socially right.

But while the nature and rise of ethical emotion is often untruly connected with some one kind of act, as obedience, or with some one kind of motive, as love of reward, a far more likely field of investigation is opened by those who connect feeling of duty with conflict of motives. Yet it is obvious at first sight that mere opposition of any two psychic factors is not a distinct feeling. I have seen my dog run away from me to follow some canine friend, and then back to follow, and so on, till one affection became 334dominant force; but such simple interference of emotions does not constitute any third and new or higher emotion. Conflicts of this sort in higher natures have sometimes a reflex psychosis in painful feeling of distraction and bewilderment, but this is the end of the natural course of feeling conflicts.

There are, however, higher phases of conflict of motives which may bring us nearer to ethical emotion. A burglar, the evening he is to crack a safe, is inclined to indulge in several glasses of wine, but his companion remarks that he ought not to drink if he expects to do the job. Here is something to be done, a duty, and under the compulsive force of the feeling of this duty the burglar lays down his glass untouched. Is not the psychic phenomenon really a case of the ethical emotion as involved in the thwarting of present inclination for the right carrying out of the thing to be done? A feeling for that which is laid upon us to be done, whether we lay it upon ourselves, or it is laid upon us by others, has certainly the compulsory quality which we commonly attribute to the ethical emotion. When we have set out to do something, this pre-determination exercises a peculiar pressure when some diverse inclination enters, but it is the force of firmly-formed purpose and of tenacious will. Its compulsiveness is not ethical, but volitional. A very little reflection convinces me that something to be done, and something which ought to be done, incite distinct emotions. I feel differently when I go to church, because I have planned to go, or have been told to go, and when I go simply because I feel I ought. There is also superadded, the purely impulsive force of the emotion for the larger good; and this may, indeed, play the whole part in the contest with present inclination, which contest then becomes of the simple alternating order. Thus the burglar has avaricious visions of gold, and relaxes his cup; he looks at the tempting wine, and grasps it again, and so on.

335It is true, however, that the feeling for the larger and future good against a present inclination may be a feeling of oughtness, a feeling of duty, a constraining to do a set something. Providence and prudential action are enforced not merely by, "I wish to get the larger good," but, "I ought to reach it." The most permanent, the greatest and completest pleasure and benefit not only incites us, but constrains us. Constraining emotion, a feeling of oughtness, may then arise both from a preview of bare accomplishment of plan or purpose set by ourselves or others, and also from sense of larger over lesser advantage. Here is the region of utilitarian duty, of the Ethics of calculation of personal pleasure and happiness. Psychically here is a true feeling of ought, and here is the ethical emotion, if we make the term denominate all feeling of oughtness. But if this is the region of Ethics, it may be said to be the region of the lower Ethics, and we may indeed deny the term ethical to all this kind of emotion of oughtness. The emotion arises about personal and particular ends, and not about principles. The ambitious man feels an ought as well as the conscientious, but they are diverse in nature. Alike merely in the general quality of compulsive force, they may differ in tone and special qualities. The constraining emotion which comes with viewing a universal law of right may be claimed as distinct from the constraint exercised by personal ends. But it is not our purpose to discuss this matter here.

The psychic conflict which is specially connected with moral emotion is the conflict of the egoistic and altruistic impulses. When in such a struggle sympathy prevails, we approve as objectively moral and right, but the existence of ethical emotion in determining the dominance of the altruism is not assured. Pity originally overcame hatred without the compulsion of duty. Altruistic impulses contest with egoistic in naïve and simple natures without any appearance of feeling for duty. The origin and 336nature of morality does not thus seem bound up with the earliest forms of egoistic-altruistic contests, though in later evolution it may come in as reinforcement of the altruistic. We may feel then, not merely like helping a man in distress at the expense of our own comfort, but we feel we ought to help him; the force of a general principle of conduct is felt in the form we term the ethical emotion, yet it is obvious that such a recognition of a general and universal law and such a feeling therefor is far later than the rise of altruism itself. Darwin alludes to the baulking of the social instinct as having special ethical significance. With the social instinct baulked, as with any other, there certainly results distress, but it is by no means made clear that this necessarily involves moral quality. When a savage in a fit of anger slays his pet child, the misery of baulked parental instinct may soon be felt, and he may bitterly regret the deed, but this does not involve moral feeling, a feeling of repentance for the essential wrongfulness of the act. He would regret in the same spirit the destroying his dinner by his own hand. If we say that he is stricken with remorse, we assert conscience violated. Remorse cannot explain conscience, but must be explained by it. Still, morality is not bound up necessarily with sociality. Sociality certainly arises and progresses to a considerable evolution before moral compulsion and the emotion of bare rightness arises to sanction and to stimulate social activities. And if moral emotion is not implied positively in altruism as an outgoing towards others, neither is it implied in the incoming of others upon the individual, either in respect of approbation or disapprobation, or in the more direct and essential way of rewards and penalties. Penalty is at bottom but a species of disadvantage brought to bear on the individual through fear of consequences. The desire to get even—an eye for an eye and a tooth for a tooth—and all exacting justice as an equivalence, whether as exacted by the individual or by 337persons delegated, the officers of justice, is plainly not in its origin and basis the ethical emotion. A system of mutual dues and rights may or may not have the sanction of morality, but they arise in advantage; and the motives which originate penalties and act with reference thereto, are far from being the pure moral emotion, a direct feeling for rightness as rightness. The merchant in general pays his import *duty*, not as a moral duty, but as something required by legality rather than morality. Law and public sentiment exercise through emotion, and that of a compulsory type, certain effects on conduct, but it is clear that the general feeling of oughtness as self-imposed law of rightness is not presupposed.

If the ethical emotion be not specially bound up with obedience or with conflict of motives, may it not be particularly connected with science? At the outset we note that a very natural confusion of science and Ethics is favoured by the fact that we can apply the term Ethics both to the science and the matter treated, and so speak of the science Ethics as the science of Ethics, of ethical perception, emotion and action. But yet we know that the science is by no means to be identified with its subject matter, and also that the science of a matter and the Ethics of it are two very diverse psychic tendencies and points of view. Science is always an objectifying impulse whose end is merely to know, but Ethics is subjective, whose end is merely to be. This is emphasized by the fact that science in its ceaseless objectifying may constitute a science of science, and science of the science of science, and so on, but Ethics is self-contained, and there can be no Ethics of Ethics. While we so sharply distinguish scientific and ethical activity, yet so far as the science is prompted by ethical emotion it is ethical activity. If I learn and know out of the feeling of duty, the act is psychically moral, yet is always distinct in quality from the feeling 338which prompts it. Thus there is an Ethics of science, or rather, to or toward science, though most scientific activity is carried on at the stimulus of other impulses, as love of truth, ambition, etc. Psychologically speaking, then, science is in no wise Ethics nor Ethics science.

But it will be said, "Is not ethical discrimination a cognitive activity? Must not one know the right, know that he ought, before he can feel ethically and act ethically?" But it will be found that at bottom the rightness of an action is the appreciated accord of the action with an end which is already felt to be right. I am asked whether I think it was right for a certain poor man to purloin a loaf from a baker for his starving family. In passing ethical judgment I simply fall back on some ethical postulate. The right of the family to life, I may say, ought to take precedence of the right of property. I therein fall back upon the simple feeling of right as ethical emotion. The knowing activity is concerned merely in the apprehending the situation, and ratiocination in tracing back to moral principles, but the ethical discrimination is neither, but an affair of direct emotion. If it be felt to be right to save life in any wise that seems necessary, I will approve it as right. A reason can only make an act right by being a right reason. Thus it is that moral discrimination is at bottom no more than a peculiar feeling about acts, towards or against the doing them, which, like all emotion, involves the knowing its object, but is not involved or explained in its psychic quality by the knowing act. The setting out what ought to be done, the establishing duties and moral rules of conduct, the development of a system of Ethics, is not then fundamentally cognitive process, but emotive. Hence it is, psychically speaking, a misnomer to denote any system of Ethics a science.

It is true we may denote by Ethics—always capitalizing the term—that branch of psychology and sociology which investigates the nature and laws of ethical phenomena. 339This Ethics merely gives an objective account of ethical emotion and conduct. It is often defined as the science of conduct, a definition quite too wide, for conduct is action consciously self-directed to an end, be the impulse anger, fear, love, ethical emotion, or any other emotion; but psychological Ethics studies only conduct as moved by ethical emotion. Conduct is, indeed, the sphere for ethical feeling, and any specimen of conduct, whatever its psychic stimulus, may excite moral approval or disapproval and stir ethical emotion, but this ethical survey of conduct is not properly a science, as has just been shown. All conduct is then objectively interpretable as moral, though it be inherently and psychologically immoral, that is, having no element of moral feeling. The spheres of objective and subjective morality are far from being coincident.

Further, science is not peculiarly related above common knowledge to ethical emotion. Common sense and ordinary fear lead me to jump off the track before an approaching train, while physiological knowledge and ordinary fear may incite me to put on rubbers on a wet day. Scientific knowledge opens the way for the common emotions; it shows the consequences of acts with fulness and accuracy, and so opens a wide range for the ordinary emotions which awake at sight of the experienced and experienceable. If I feel I ought to put on rubbers, this feeling arises, not directly at the consequences which science reveals, but at the rightness of the consequences. I feel I ought not to injure my health, a feeling which science does not generate, but it merely establishes the fact that such and such actions will injure my health, and so gives the opportunity of applying the moral postulate, I ought not to injure my health. I judge the rightness of an act, not by its consequences, but by the rightness of its consequences.

Again, science reveals most clearly the necessary means to ends; it says that to make nitro-glycerine you must 340use such and such ingredients. In viewing these means in their necessity there may arise a certain emotion of compulsion to their use; but this compulsive quality is not, I ought to do so and so, but I must, if I would attain the end. It is plainly an unethical use of terms to say, If you wish to succeed or be happy you ought to do so and so, or that is the right way to succeed or be happy. Morality is not a recipe toward any end but itself. So the feeling as to the "Conditions by fulfilment of which happiness is achieved"— emphasized by Spencer in the principles of Ethics as the main element in moral emotion—is not real ethical emotion. I may feel the constraint and necessity to using certain means, difficult and unpleasant in themselves, in order to reach a desired end, but a moment's introspection shows that this compulsive emotion is not thereby moral, that this feeling is not a feeling of duty but of necessity to employ the means. If I feel that I ought to become happy, then alone will I feel I ought to use the means to happiness. So also a man may desire to win in athletic competition, but the requisite means, a hard course of training, may deter him from entering; that is, his love of ease conflicts and overcomes his desire of athletic success as far as action is concerned. If he undertakes the training and struggles through, he feels the compulsion of the means in direct proportion to his love of ease and pleasure. He refuses a cigar under this emotion at the necessity of the means, but this is plainly not a case of ethical emotion; he refuses, not because he ought, but because he must, and the trainer who says to him, "You ought not to take that cigar," does not primarily appeal to moral principle, but to the constraint of the means to desired end. This does not deny that a man may feel training as a matter of duty, but it is still obvious that he who refuses a cigar as a mere matter of training, is as psychic fact actuated by an emotion of distinct quality from that which the man feels who refuses 341to smoke as a matter of conscience; the feeling, "I must not," is diverse from the feeling implied in, "I ought not." The athlete may be conscientiously an athlete, but in general he refuses to smoke merely because that is the right stand, *i.e.*, suitable to gaining the particular desired end, whereas the conscientious man refuses as determined by a feeling for some end whose rightness is assumed, as the preservation of health, or the being inoffensive to others. The athlete is moved by what is right or useful to some end, while the psychically moral man is actuated by the emotion for the end of rightness; and while constraint appears as characteristic of both emotions, still in breadth, depth, and particular tone, the ethical is plainly differentiated from the necessitarian emotion. At bottom also it is plain that the feeling of compulsion to means is a case of conflict of motives—as with the athlete is love of pleasure of smoking *versus* desire of athletic success—and conflict of motives has been previously discussed.

Neither scientific nor common knowledge then can as method of means give by itself the moral emotion. But it may be said that science does provide ends for action and that the emotion about the end is an ethical emotion. Thus the end of truth, of adherence to reality, is naturally emphasized by science; yet here is not duty, but the essential guiding emotion is the emotion for achievement and the achievement of the desired accordance with nicety and completeness. The enthusiasm for truth and truth in action is an emotion which may be sanctioned by moral feeling, but it is not moral feeling. Adaptation to environment or conformity to reality as a general end of action may have its impetus in moral emotion, I may feel that I ought to accord with the nature of things as scientifically revealed, but this motive is by no means necessarily implied in the end. And conduct is rarely actuated by pure sentiment for this end; rather the general form is, "Do this and thou shalt live"; that is, the emotion is 342desire for personal ends to which accordance with nature is the means.

Again, take a suggestion of end for conduct from some special science. For instance, Biology marks as the general result of the struggle for existence and of natural selection, the perfection—practical and relative—of the kind. Thus the result, that is, end unconsciously achieved, of the life of deer is power of locomotion and keenness of scent, while with man the tendency of evolution is toward brain power. Man obviously is able to consciously make an evolution tendency an end, to conduct himself with reference to it, and thus man's life may be a conscious and strenuous carrying out of tendency. A constraint arises from this end as from others, but it is not moral constraint, till the end has been adjudged right; thus this end does not explain rightness. The aspiration toward self-culture and self-fulfilment is not psychically moral, nor yet the determination to achieve this perfection. Perfection, be it remarked, is not an end, but the measure of attainment of any end; a perfect man is one who is complete in certain respects. Morality is not the carrying out any end, perfectly or imperfectly, be it pleasing, satisfactory, true, good, etc., but it pursues and is pursued by the right end, which is rightness as universal, authoritative, compulsive, self-approved, impersonal law. The emotion of oughtness in its purely ethical form is responsive to this alone. Purely moral emotion as psychic fact, is not any feeling for any *summum bonum* or any perfection of attainment of any kind, but is an emotion for the right for its own sake. It is neglectful of all consequences, and cries, "Let justice be done, though the heavens fall." We all know the distinct difference in quality of feeling when acting merely to do my duty and when acting to achieve an end for the achievement's sake or for the good implied. Ethical emotion may arise about any extrinsic end, but does not arise out of it.

343We conclude then that as psychical fact there is a variety of compulsory emotions, an ought of law as behest of others, an ought of means, an ought of end, an ought of advantage, an ought of bare moral rightness, and that this latter emotion, as every one knows by introspection, has its own peculiar quality and force. He who feels constraint from authority, from use of means, from end purposed, is plainly feeling different from him who feels the constraining emotion at moral right. And the law which says, "Do this and thou shalt live," does not bring moral pressure, for the moral law says, "Do this whether thou livest or not"; that is, moral emotion and activity is not consciously to itself a life factor. As a matter of psychic fact a world of moral activity exists solely for and in itself, and the emotion in this sphere of absolute morality, in which many conscientious people live habitually, is ethical emotion in the narrow and strict sense of the term. The immediate feeling of absolute rightness—so-called intuitive morality—however and whenever it has arisen, seems to present itself as mental factor radically diverse from all emotions of means, ends, and law.

Here we may criticise a so-called rule of moral conduct to which appeal is often made, namely, the rule that we ought to do as we would be done by. We know, indeed, that the principle of equivalence is strong in society, and that if we wish to be well treated we should treat others well. However, to do as we would be done by, in order that we may be done by as we would, transforms moral precept into prudential maxim. Here is a method of advantage: in order to attain the given end we ought to do so and so, but the purely ethical emotion is not aroused. But further, interpret the rule as simple universal moral law that we ought to do as we would be done by. This involves putting ourselves in another's place and considering how we would like to be treated under the circumstances, and so treating him. This is hedonistic altruism, 344and its measure is crude and unreliable, for what might please me in a given case might not please another. This automorphic interpretation is, however, extremely common, especially in lower psychism. The child and the savage judge inevitably and naturally that they are giving you the greatest pleasure when they share their dainties with you. But slowly is individuality of taste recognised, and still more slowly recognised as proper and right. Still a hedonistic altruism, whether by mistaken mode of putting yourself in his place, or by true measure of realizing what he is in his own place and acting accordingly, on either method is of very doubtful morality if judged by any high standard. Indeed, hedonistic altruism, whatever its motive, has wrought both incalculable injury and unrighteousness, whether as a weak sentimentalism as seen, for instance, in promiscuous charity, or in more special forms, like parental indulgence. Ethical emotion which seeks to be directed in its action by an extraneous measure adulterates itself. We ought not to do to others as we would like them to do by us, nor yet as they would like, nor yet merely as we feel they ought to be treated, but the real golden rule is, we ought to do by others as we feel that they in their own nature and position ought to be done by. This is no more than to say that we ought to do by others as we ought, a moral identical proposition; and the reducing to this shows that moral emotion rests only on itself. The end of pure ethical conduct is always and ever merely to fulfil righteousness everywhere or to secure its fulfilment everywhere, to help and forward all doing right. The so-called golden rule may have its place, as undoubtedly it was meant, as propædeutic to a kingdom of righteousness, but it has not pure ethical quality in itself.

CHAPTER XX

THE EXPRESSION OF FEELING

The primary function of mentality, as we have throughout assumed, is as stimulant to activities advantageous to the individual under the conditions of its existence; hence all these activities are in a broad sense expressions of mental state, they are the outflow of psychoses and are indicative of them. In particular, feeling is specially and directly related to motor values, which thus become to the self-observant or to others observant an index or expression of the feeling. Thus, I see a deer fleeing from a wolf, and I infer that this is an expression of fear. Hence we may rightly say that in a large sense all action is expression, for all such action rises in feeling; in other words, from one point of view expression equals action. Not only may exterior bodily phenomena betray the feeling which is their inciting cause, but to a vivisectionist, for example, interior phenomena, cerebral and other, may be noted as indicating a feeling origin. Excluding, of course, so-called reflex action, which is really reflex motion, action and expression are but different points of view of the same thing: what we term an action when we dwell upon the motor side, we term an expression when we dwell on the mental *prius* and stimulus which is revealed.

Now as the evolution of mind progresses actions no longer serviceable may survive in connection with given feelings, remain indicative of them; thus the strong beating 346of the heart in fear and the scowl in anger. Such survival actions which occur in connection with all kinds of feelings, and especially with those which are pre-human in their origin, are with particular emphasis styled expressions. The scowl in anger is considered as expression rather than the actual blow struck, which is equally the result and indication of anger.[F]

F. Wundt says that when in emotion we look "sour" we think we are actually tasting the sour, and so make the repulsing action, "sour" look. (*Lectures on Psychology*, p. 283.) I think it more probable that the "sour" look is the survival expression of such an emotion as disappointment. It is likely that the genesis of disappointment was in tasting the sour for the supposedly sweet, *e.g.*, lemon for orange, and the "sour" look has remained as expression of disappointment long since its utility ceased. The genesis and early growth of most emotions is in connection with certain sense experiences and their related actions, and these actions tend to remain as "expressions" long after their real quality as actions has disappeared. Hence it is by survival, and not because he thinks himself tasting something sour, that a man looks "soured" by disappointment when I fail to give him money as promised. So also black is gloomy because we are diurnal, and our ancestors were diurnal. If nocturnal, black would seem joyous, white gloomy. (Cf. Wundt, *ibid.*, p. 375.)

Expression is then primarily all action connected with all consciousness, secondarily, it is useless action continued by force of habit and transmitted to descendants. But still many expressions are more than mere actions or their survivals. To be sure, Darwin and many Darwinists maintain that the expressions do not arise or exist for their own value as such, but they are entirely incidental. Expression is not the function of the so-called expressions, but they are entirely functional survivals. While, however, we must admit that many expressions have arisen and been preserved in this manner, yet I think it is altogether hasty to deny the function and value of expression *per se*. Expression has existed as a function from very early phases of life, and it underlies all bisexuality and sociality which have been such important 347elements in evolution. Organic sound-producing structures, whose sole utility from the very first is for attracting attention, early appear, and further voice seems to have its origin in the demand for love-call and call to young. Gregariousness is made possible in almost all its forms by purposive expression. There comes early, then, a will, not merely in performing some definite act at prompting of a feeling, but also a use in simply expressing it to others, communicating the fact of having pain or pleasure states to others. The cry of pain in young animals is a cry for help, and as such has been favoured in the struggle for existence. The usefulness of this action is solely as expression, and as expression it has arisen and been developed. Expression here is not an incidental view of a physiological action, but exists for its own value to the individual. Such expressions have their use in their significance, and as the true language of feeling are to be interpreted by the principle of serviceability. An expression which is and continues, by reason of its utility, as a sign-language, visual, auditory, or otherwise, as gesture love-calls, etc., may be termed pure expression as distinguished from incidental expression, like blushing, pallor, etc., which exist, not for their significance, though they are significant. Incidental expression includes also the sphere of degraded action. Yet what seems mere degraded action may be true expression, as beckoning, which is an abridgement of the action of pulling one to oneself and of movement towards oneself; but this motion of the hands exists, not for this end, nor as survival, but merely as significant of a desire on the part of the gesturer. In the higher ranges of life we well know the large place played by pure expression as distinguished from incidental expression. It is not necessary to suppose that pure expression consists merely in "voluntary and consciously" employing "means of communication" (Darwin, *Expression of the Emotions*, p. 256); thus, the scream of 348an infant is equally pure expression, whether the infant employs it knowingly or not as such, for screaming of the young has doubtless arisen and been preserved in natural selection because of its utility as significant. There is then, I think, a group of activities which are not merely incidentally expressive, but originate and exist for expression as a useful thing in the battle of life.

But we have not exhausted the principles of expression when we refer to present or past serviceability as an action in general or to service as expression. It is plain that in any activity prompted by any feeling there comes at a certain high intensity a more or less pathologic over-functioning of the organs concerned, with under-functioning of others. Emotion as action stimulator in any high degree always enhances some physiologic function to the depression of others. The blood, for instance, is forcibly withdrawn from various parts to certain specially active parts, and this withdrawal gives rise to an appearance which may be termed a negative expression, as the pallor in fear. Certain other phenomena connected with fear, as change of colour in the hair, cold sweat, and trembling of the muscles, which are mentioned by Darwin as unexplained, are probably due to this negative principle (*Expression of the Emotions in Man and Animals*, New York, 1886, p. 350; but compare pp. 81 and 308, where these disturbances are ascribed to direct action of the nervous system. Darwin does not, however, distinctly state or treat the principle we here mention as a distinct law). As the body is an inter-related system of organs, stimulation to one organ means an effect upon all, excitation of some, depression of others; thus to an acute observer the whole body is symptomatic of every feeling, and, indeed, of every consciousness. In the natural and normal course emotion, to do its work most effectively, implies little or no marked negative expression, but the nervous energy generated flows freely and directly to the organs which 349are to do service, without greatly impairing general function. Fear thus acts at first simply and advantageously; but in its later history fear becomes greatly complicated, and instead of freely issuing in serviceable action with not excessive heightening or depressing of any function, its outlet seems as it were choked, and the nervous energy spends itself within the body in violent disturbances of vital organs. Fear becomes then decadent and loses its place as evolutionary factor, becoming impediment rather than aid to progress. Negative expression must then be considered as especially notable in the later exhibitions of an emotion when concentration becomes morbid and ineffective, losing its advantageousness, and the emotion is being supplanted by other psychic factors. Great injury and death itself may result from the abnormal action of fear and other primarily useful psychoses.

Besides the particular organs to serviceable activity with the subsidiary physiological functioning, and the indirect depression, we must still note other principles which may control expression. Nervous energy under the incitement of emotion is often in excess of the demand for the required action, and it will then overflow into correlated activities along the line of least resistance. Also when the suitable action is checked for any reason, its motive force backs up and overflows in new channels. Indefinite and purposeless movements of various kinds thus result which may be expressive of the emotion of which they are incidentally the result. Any one who has watched an Irish setter tracing game must have remarked the wavings of the tail becoming more rapid when the scent becomes stronger. When the dog is running very fast, the tail-waggings are less noticeable than when moving slowly, although the interest may seemingly be the same in both cases. It is obvious that a fast run uses to a large extent the superfluous energy which was discharging in tail movements, and when the useful running is 350checked the tail motion recommences with greater force, serving as a safety-valve. The frisking of young animals and children is also largely due to diffusion of so-called superfluous nerve-force, and is expressive of general sensations of pleasure. All feeling is motor in its natural value and tendency, and unless the resulting energy is fully used in some special serviceable action, it will discharge itself along the easiest and most habitual lines laid down by inheritance. Thus the peculiar ancestral experience of animals is always expressed by their spontaneous diffusive activities. It will be remarked that the principle of diffusion is the reverse of negative expression, being an overflow of force as opposed to withdrawal. Excessive generation of energy is certainly uneconomical, and we must consider that at first emotion tended rather to less than the required amount, than more.

The phenomena of diffusive expression, in the strict sense, are thus rather late in appearance. The very lowest forms of life have no infancy or play period, and from the first are directly active in the struggle for existence. Yet the play period was certainly evolved through natural selection as a fully educative and preparatory stage, wherein the actions most demanded in actual life are unconsciously practised and a general basis of reserve force is accumulated. Play activity is a living on inherited energy and in the inherited modes: the kitten pouncing, the horse prancing, etc. Play is then rather a mode of activity than a mode of expression; it is expressive only in the way that all action is expressive. Expression proper is only in those modes of action which are carried on, whether consciously or unconsciously, by virtue of their significance value. If everything which is expressive is called an expression, we must include all the bodily actions and phenomena which can in any wise be connected with consciousness. I use the term diffusion in the narrow sense of spontaneous overflow of energy in 351excess of that absolutely required for the advantageous action. I do not refer to the general diffusion of emotional effect throughout the whole organism, which always occurs by the very nature of organism. Thus the pain from a pin-prick certainly modifies to some extent every cell in the body; there is a direct wave of influence from the psychic experience, and this is propagated throughout the whole organism by reason of its essential interdependency of parts; it echoes and re-echoes throughout the whole. The physiological result is then in simplest cases extremely complicated. However, this mere general fact of diffusion is a biological truism, and does not explain any expression, but simply asserts that every feeling, by virtue of its physical basis, affects the organism as a whole. Emotion issues specially in motor activities because its origin was as stimulant to necessary action, but this action involved internal organs, especially the circulatory and respiratory, and indirectly the whole body in every part. The explanation of an expression must always be in tracing back to the original serviceable actions with their demands on special subsidiary organs, and their depression of certain related organs, and not in reference to the general law of diffusion, which is but another term for the essential continuity of the organism. A useful principle of expression must not merely say that there is by the nature of organism a general bodily result from every emotion, but it must explain the particular expressions.

We make them so far four principles or forms of expression, which we instance in saying that the blow of an angry man is general activity expression, shaking the fist at one, purposive expression, scowling as remnant of watching foe intently in the open air is survival expression, and twitching and trembling of certain muscles is diffusive expression. Every emotion commonly issues in all four forms, in direct activity with associated survival tendencies and purposive expression, and a surplus of energy runs 352over into certain natural and easy motions, or a deficiency of energy in certain organs manifests itself, the negative side of diffusive expression.[G]

G. Since emotion comes in waves, expression is reduplicated. This may throw some light on such an expression as laughter. Landor says the Ainu do not in the proper sense laugh, but they roar with delight. It may be that laughter is reiterated roar as resulting from reiterated psychic impulses and feelings. As in the growth of an emotion, waves are multiplied, the expression becomes more reduplicate, and thus laughter tends to become more rippling and articulate. The cachinnation and explosiveness has thus a plausible explanation, which I merely suggest. At least Prof. Dewey's explanation (*Psychological Review*, I., 559) that "both crying and laughing fall under the same principle of action—the termination of a period of effort"—is quite too general. Tension ceasing, effort stopped, we "breathe freely," take deep inspirations. Laughter is far from being the usual outcome of such a *status*.

Darwin makes antithesis a principle of expression. Thus the expression of affection in the dog or cat toward its master cannot, says Darwin, be traced in any wise to serviceability, and we must seek its explanation merely as unconsciously and instinctively assumed as directly contrary to the serviceable hostile expressions. A dog's expression of anger is, or has been, directly serviceable action, but the expressions of affectionate pleasure seem never to have had such an origin, but to have arisen merely as antithetic to the former, and so establishing the utmost distinctness of impression. To convey most clearly a motion of its friendliness the dog naturally assumes those attitudes which are most diverse from its expression of hostility. Their serviceability as expressions is best attained by being completely antithetical, and the more antithetical the better under natural selection. However, if this be the case, antithesis scarcely deserves, it seems to me, the name of a principle of expression, but it merely denominates the fact that opposite emotions in the struggle for existence tend to exhibit themselves in opposite ways as similar emotions in similar ways; but 353we need neither antithesis nor similarity as a principle. I believe that serviceability past or present either as direct action or as expression is the prime *impetus* of what we term the expression of the emotions, and I confess I do not see much force in Darwin's Chapter on Antithesis. If, however, opposition has a meaning for life, as Darwin seems to imply, then does not the expression come under the law of serviceability? If there is any opposition in expression, I should explain this in general by utility rather than by antithesis *per se*. Thus take the gestures instanced by Darwin (*ibid.*, p. 65), of pushing away with the hand when telling one to go away, and of pulling toward when telling one to come; these gestures are, indeed, antithetic, but their explanation does not lie in the fact of the antithesis, but in the fact of the past serviceable habit, by which individuals disliked or liked were repelled or attracted. In the present instance the person motioned to may be far beyond the reach of the arms, but still the gesture may be more than mere useless survival, for it acts as emphasis of the vocal expression, and has its influence there.

Darwin for some reason constantly ignores the serviceability of expression as such—not so much as a fact, but as a principle—and hence its relation to natural selection, whereby he involves himself in needless difficulties. If an expression is of use, why should it not arise through natural selection as well as a limb, a wing, or an eye? Like other functions, expression may be incidental or may adapt variations attained originally for other ends, but in the case of the voice, at least, we have an original organ of expression as instrument of intercommunication.

Nor can I think Darwin's treatment of the expressions of affection by the dog as due to antithesis a very happy or satisfactory solution. In the first place, the expression of friendliness by the dog is not the complete antithesis of that of hostility. The dog barks both out of friendly 354joy and from anger, as Darwin himself states. Some dogs also, as I have often observed in my dog, show pleasurable affection by wrinkling up the lips and showing the teeth, an act which is often mistaken for a hostile demonstration. Dogs also, as is the habit with my own, will often express affection in the same way as the cat (*ibid.*, fig. 10), by rubbing against one. This is but an instance of a general law of expression of affectionate emotion, *i.e.*, closeness of contact with the beloved object which is liked as promoting pleasure. This instinctive expression of love or liking certainly had its origin in serviceability, the appropriate effort toward the pleasure-giving thing or animal, but specially in the relation of parent and offspring, and in that of alliance in danger. Again, the tail of a hostile dog is, as figured by Darwin, straight and erect, but the opposite of this is the tail tucked between the legs when fleeing from pursuers in fear, rather than the position when showing friendliness to its master. My own opinion of the rise of the friendly expression of dog, cat, and other animals toward man is that they are in the main, at least, transferred from the serviceable friendly expressions used among themselves in a wild or domesticated state. I have repeatedly seen small dogs, who attach themselves to some large dog as their master, fawn, posture, and lick this master precisely as this master does his human master. Dogs and cats also show their affection and care for their offspring in many expressive acts which are transferred to their human owners. These expressions were primarily either directly serviceable actions, as the licking, or serviceable for expression as such, as various sounds made to give assurance of presence of food, or of safety. In general, it seems to me that when antithesis has occurred, it has arisen out of serviceability and not *vice versâ*.

With reference to the wagging of the tail in the dog, this is far from being an expression of affection alone. I have already mentioned the case of the setter where the 355movement of the tail is largely due to diffusion of superfluous energy, analogous to nervous habits like pacing the floor or biting the nails in human beings. With some dogs at least, as I have noticed in my own St. Bernard, the tail is switched slowly back and forth when approaching another dog with hostile intent. We have not as yet a sufficient number of facts at hand with reference to the history of the dog to pronounce the tail wagging as originating by virtue of its use as expression. And what is the *rationale* of the origin of the tail in the dog and cat, and for what reason has it been perpetuated? Is it a prehensile survival—which has been taken advantage of in the breeding of the pug—or is it a sexual characteristic, or did it originate to perform some directly advantageous action, as the tail of the cow and horse, or did it come into being as an organ of expression? Is the tail-wagging recognised by animals themselves as an expression as it is by man? These are questions on which we must have more data than we now possess in order to make any sufficient answers.

Again, the rise of the barking by dogs under domestication is another problem on which little can be said with certainty for lack of data. Darwin's remark that it may arise by imitation of the loquacity of man seems to me ludicrously inadequate, and there seems no element of imitation in the noise produced. Domesticated animals in general tend to use the vocal organs for louder sounds than when in the wild state, for with wild animals the value of a loud noise as expression in any way is largely counterbalanced by its betraying presence to enemies. When natural enemies of the dog are driven out by man there will be a tendency toward a larger use of the vocal organs, both with reference to companion dogs and also to man. The particular sound, the bark, is determined by the nature of the whole vocal apparatus. The bark was, no doubt, originally to frighten aggressors, 356as I have often seen a large dog frighten a small dog from a piece of meat by a sudden resounding bark. Gradually attained as a mode of terrifying his competitive associates and certain game which it follows under domestication, and so preserved and developed by natural selection, the tendency is also powerfully strengthened by artificial selection, the best barker, other things being equal, being chosen for breeding by man. When the bark has become a common and habitual practice, it becomes a vent for superfluous energy developed by joy and other emotions. Like snarling or grinning, it is also a play form, and thus becomes denotative of joy by association. To impress one's friendliness or hostility upon others, to appease or terrify, are the two main ends of expression with both man and animals, and this function is excited in various ways by different species, as determined by environment. The danger signal and the safety signal, the beware or welcome, is amplified and varied according to particular requirements which must be fully investigated before we can give any complete *rationale* of any expression. Conciliatory and menacing expressions and gestures have been evolved and matured in strict correlation under the same general law of natural selection, and neither one nor the other is due to antithesis. It is entirely unlikely that of such expressions, one, the hostility side, was first developed by natural selection, the other owing its rise to a distinct principle, antithesis.

However, I am not ready to deny antithesis all force as principle of expression, but it seems to me it should be ranged with law of similarity or analogy as subsidiary, and largely influential only in the higher types of expression, especially the teleologic human, as in gesture. Thus, if thumbs up means pity, thumbs down would naturally be used to denote pitilessness. To nod the head means assent or yes, to shake the head means dissent or no, though the exact antithesis would be to throw the head 357backward—assent signal with some tribes. However, while it may be asserted that, as a general law, that like emotions express themselves in like ways, unlike in unlike, this can hardly be used to throw much light on expression. Given a particular emotion and its expression, we can by no means deduce immediately the expression of the opposition emotion. Particular conditions and special organic limitations will always make this impracticable, and it is the office of the scientist to study expression in the course of evolution as of service under a multitude of conflicting interests and distracting difficulties.

We view expression then as mainly due to the principle of advantageous variation in the struggle for existence. Expression is the action required in the battle for life, or accompaniments to assist this action, or the call for aid to bring it about. Natural selection is the first and fundamental law of expression, negative expression and superfluous energy both being secondary and often pathologic in tendency.

The struggle for existence is itself on the very face of it an expression of mind, namely, activity significant of certain will and feeling experience. Whatever shows mind is expression, and thus in a large sense every movement in the physical universe—and what is the universe but motion—and every organic activity may be construed as expression. Whether all force, motion, action, is or must be expression is, however, a philosophic investigation which we need not now discuss, though we may suspect that the height and depth of mind and so the range of expression is enormously beyond the science of to-day. However, restricting ourselves to the domain of animal life, it is obviously very difficult to determine just what activities of a given organism betray mind, and still more just what form of mentality is manifesting itself. Man, being the measure of all things, interprets himself in all, 358and even when he becomes aware of the dangers of anthropomorphism he cannot wholly disengage himself from the tendency. The subjective analogical interpretation is a necessary evil. Still man is the keenest sighted of all beings for expression, and actions in a very wide range which had not in fact the real function of expression become expressive to him. The primary value of fear for the deer is to make it run from danger, and the running becomes expressive of the fear to observers, though the running is not for the expression. Thus vital activities of many kinds are expressive, though their primary value is not in the expression. Activities whose sole or main value is to give expression are comparatively late, the value of expression in this narrow sense being in the mental impression thereby made upon other organisms. Thus actions which serve purely to frighten others, in making one's self formidable by loud noises, as roar of lion, bark of dog, by erecting the hair, displaying claws, teeth, and other such actions are pure expressions.

There is a constant growth in the value of expressiveness as we ascend the scale of life, expression playing a larger and larger part till with man certain individuals become specialized as expressionists, artists, poets, and orators. Further, fine art is expression which has its value, not in any exterior utility, but in itself alone, the subjective emotion seeking in a manner perfectly free from the common utilities of life to find itself a complete and perfect embodiment. Art here does not serve life, but life, art. The experience has in itself its own vindication for being, in that it expresses. Expression is no longer bare action nor yet a function to serve life, but it becomes a life in itself. In this ideal life of pure expression we recognise the necessity that the expressionist be emancipated from the struggle for existence, be freed from the sordid cares of life, and given up wholly to expressing his individuality with characteristic force; hence the State often pensions 359writers and artists. But apart from this ideal life, in the evolution of intricate sociality and industry and complex culture, expression becomes a more and more potent factor. Man in society must not only be, he must reveal himself, he must show what he is in order to achieve the most. Many fail, not for lack of faculty, but for lack of expressive ability. Expression, then, in general, is a function which, starting from the most minute beginnings in the lower animals, culminates in man. In large part man is man by reason of his superior power of expression, especially by speech, oral and written. Evolution in man is on the mental side in particular, but a large part of this mentality has been given to the improvement of expression in making it more facile, full and rapid. The complete natural history of expression is yet to be written, and all that I attempt is to indicate the point of view for such an investigation.

There are two points further with reference to expression which merit a few remarks. The first is as to the reaction of expression on emotion. We have treated to some extent the relation of emotion to its expression, but we have also to consider the relation of an expression to its emotion and to emotion in general. We have all along assumed that the emotion as a factor in the evolution of life is an internal stimulus to a serviceable activity, which may be viewed as its expression, or may even have its value as such. That emotion, as stimulus of action, determines expression is, I think, a primary law. However, Prof. James maintains (*Mind*, xxxiv., 188) the reverse—that expression determines the emotion.[H] We do not strike because we are angry, but we are angry because we strike. Hence, in reality, the emotion is really the expression, that is, the emotion is the consciousness result of the so-called expression—it expresses the "expression" in terms of 360consciousness. We commonly speak of expressing our emotions, but we should rather speak of emotions as expressions in consciousness of certain bodily activities. But if we make emotion but a psychological incident and off-shoot of certain activities, I take it we run directly counter to the general function of mind in evolution as internal stimulant to useful activity. Emotion is, I judge, fundamentally a motive force and has its function, and so its rise and development as such. It is more than a by-product, but even if it were, how should we account for it? After the serviceable activity has actually been brought about, after a man has really struck down his adversary, what is the utility of emotion? We take it that the value of emotion lies in starting and supporting the activity, and it is advantageous economy that it cease immediately on the accomplishment of its end. While we must always suppose that emotion has its physical support in central neutral changes, yet the expression is truly such; that is, it is from a different impulse as determined by the emotional brain excitement. In the light merely of a theory of natural selection, mind in general, and emotion in particular, is more than incidental concomitant of physical changes, more than echo of corporeality: it has a vital and central function in the evolution of life. Prof. James points to the fact that exercising the expressions or imagining the feeling calls up the feeling, as a proof of his theory. This, however, is merely a matter of association, and can prove neither a real precedent nor resultant. We may call up ideation as well as emotion by producing associated activities. In the interdependence of the conscious life, emotion, perception and willing call up each other without reference to causative order. Any one element of consciousness may be regarded either as resultant or stimulant, according as we look at preceding or following state of consciousness. In the order of evolution, pain and pleasure arise from certain actions in order to inhibit or stimulate 361repetition of actions. Feeling is then, both resultant and stimulant. The emotions may arise from the expressions by association, but the original dependence is that of expression on emotion. The further test, that we cannot imagine an emotion without bringing in bodily presentation, is simply a necessity of the imaging faculty as such, an image by its very nature being concrete.

H. Professor James has of late largely modified his view (see *Psychological Review*, Sept., 1894).

While, then, I believe that emotion is the spring of expression, I am far from denying that the expression may not react upon the emotion. Whenever the will in any wise controls expression we mark modifications in the feeling. In the later evolution of life the directing of expression is of great importance, and expression is gradually subjected to the will. Hence, especially with man, it becomes possible to feel in certain ways and yet to repress the signs of feeling, to have strong emotions, and yet not betray them to those who might take advantage of them. When a strong emotion is forcibly and completely checked in its expression there is commonly rankling. At least it is not true, as Darwin states (*Ibid.*, p. 360), that "repression, as far as this is possible, of all outward signs softens our emotions." Very often, as we all know by personal experience and by observation, the checking the free expression of emotion tends to intensify, rather than soften, the emotion. The school-girl, who, on hearing sad news, rushes away to have a good cry, weeps away her grief, and experiences a deep sense of relief; while the man who sternly represses the expression of grief often suffers acutely and long. Grief, of course, sometimes lies too deep for tears, and we often long to be able to express the pent-up emotion which chokes us. This state is the opposite of the free, natural expression of feeling such as we see in children. Children express themselves without self-control, for this is beyond them; but here is the power to will expression, but the effort is always futile.

By promoting or repressing expression we do certainly 362influence emotion; but this volition is always for reason, and implies, then, a conflict of feelings. Thus, a feeling for propriety leads the man to control his tears, and this feeling in itself must tend to diminish the strength of the concomitant grief. However, though there is a measure of interference, we would be wrong in supposing that complex mental life is always comparatively weak in its component elements. The distraction of interest due to new feelings checking expression is not always equal to the relieving power of free or promoted expression. The direct checking of the expressional act certainly keeps back the current of energy from its natural channel, and the feeling has increased in duration, if not in quantity. The evanescent character of emotion with young children and with demonstrative people is well known.

But besides the changes which may come to the feeling through direct will-effected changes in the expression, we must also note that the mere consciousness of expression has often a definite influence. Thus, when greatly frightened, I may become conscious of the heart leaping into the throat, the trembling, etc.; and this consciousness of the expression acts in general as a diversion in the feeling which is expressed. Sometimes, indeed, it seems to add to the feeling, as when a girl blushes for her blushes. There is an intensification of self-consciousness which but heightens and renews the expression with renewed sense of expression, and then another flood of embarrassing self-consciousness, and so on in a long series. Here, however, the sense of expression does not in strictness add to the intensity of the original feeling, but it develops a new feeling of the same kind; at each step there is new occasion and a renewed feeling, but a total quantity is constituted, so that we are right enough in saying that the consciousness of her own blushing but added to her embarrassment. Yet it may be stated as a general law that a consciousness of our expressive acts as such tends to decrease the original 363feeling from which the expression arises, inasmuch as the field of consciousness is thereby divided.

When the will attains control over expression we may not merely repress the impulse to expression when we feel strongly, but having no feeling of a given kind we may voluntarily adopt its expression, and this adoption of the expression very often leads by association to the real feeling. Again, when experiencing a feeling we may simulate the expression of another or even opposite feeling. It is often advantageous in the struggle for existence to throw others off their guard by deceiving them as to the real emotional state; hence, craft and guile have from a tolerably early stage in evolution played a part in the history of life. Animals and men alike soon appreciate the distinction between appearance and reality, that a kind and pleasant expression is often but the lure of malice and hostility, that injury is often meant where there is the show of benefit. Plants, as well as animals, often are quite other than they appear, both for offence and defence; and there is the wide field of mimetic protection which cannot, however, at present be brought under our subject.

Simulation of expression probably arose as an economical makeshift; a mere show which costs the organism little often attains ends which would otherwise require a vast deal of mental force. Thus we see children scared into desired behaviour by assumed anger, grief, etc.; and even animals, as I have noticed with dogs, likewise frequently affect expressions which have no support in real emotion. The unsophisticated, however, learn with great rapidity to distinguish between assumed and real emotion. Any one who has made a pretence of crying before little children must have remarked this. Simulation of expression in order to easily reach desired ends is thus rather limited, but still has a real value and a considerable place under natural selection.

However, expression may sometimes be simulated in 364order to attain the associated emotion. If we act mad, we often get mad, and thus, as we see in the plays of animals and children, merely assumed expression may lead to the real emotion. This way of attaining emotion by purposely enacting its known expression, we may call impression as the reverse of the expression order. Men may work themselves up into a fury, as well as vigorously express an anger directly occasioned. Actors and public speakers often take advantage of this reaction of expression on emotion, and thereby not merely affect an emotion, but have a certain real emotion, which cannot ever be naïve. Thus Macready as Shylock used to prepare himself and get up "the proper state of white heat" by violently shaking a ladder. Poe in one of his tales makes a detective say, when wishing to know the thoughts of a wicked man, "I fashion the expression of my face as accurately as possible in accordance with the expression of his, and then wait to see what thoughts or sentiments arise in my mind or heart, as if to match or correspond with the expression." This method of acting like another, that we may have and so know his thoughts and feelings, is a very difficult way of mind-reading.

But expression is often simulated on one or both sides with full understanding of it as such. This enters into play, and is the essence of the dramatic art. That the word *play* denotes both the sportive imitative actions of animals and men, and also a dramatic representation is not fortuitous or arbitrary. It is noticeable that among the lower animals the earliest and commonest play is playing at being angry or frightened, which corroborates the view of these emotions as probably the earliest and most fundamental in life. The correlated nature of fear and anger is shown by the way they are played at; thus you often see one dog with a show of anger chasing another who simulates fear, and then the parts are exchanged.

The great relation of pursuer and pursued is constantly 365mimicked among animals with interchanging of parts. So also among children the commonest plays are those of fleeing and chasing, as tag, hare and hound, hide and seek, etc., the fundamental elements of life being re-enacted under the superfluous energy which tends to flow most easily into the oldest and most habitual channels. Thus play has a high historic psychic importance. To attack and to run away are the most necessary and essential of life activities, and play has a certain pedagogic and preparative value, and has thereby been sanctioned by natural selection, for we see that in the evolution of life the tendency has constantly been to lengthen the play period. Among the lowest animals the individual at birth is immediately thrown into the struggle for existence and must battle for itself; there is no play time for it, but at once it enters upon direct life struggle; but in higher life there is a period of spontaneous free dramatic activity.

But not only is anger and fear shammed as a prominent and primitive play, but it is most common to stimulate to real anger or fear, and then in glee to show the inadequacy of the occasion to the victim. Every one has observed how frequently young animals play by teasing and scaring each other. The tricks of boys and practical jokes of men both point to the deep inbred power of anger and fear in life, and are at the same time symptomatic of their decline in power as dominant life factors. All children delight in scaring one another on pretence, in seeing the real expression and feeling themselves the moving powers in bringing it about. This satisfaction, which is aboriginal, which is the reflex of the original pleasure sanction when power to scare others for one's own benefit was being evolved in life, makes a large part of the enjoyment of such action. A large part of play-pleasure must, indeed, be set down to reflex of the earliest hard-earned pleasure experience; but a large part is also due to the thrill of excitement and the delight in activity *per se*. Later forms of plays are largely 366due to pure imitative propensity, though often helped by reversal tendencies.

We note also that this perceived groundlessness of the action becomes a large element in later forms of play, as wit and humour, but the pleasure is plainly based on the power and the superiority of intelligence implied. It amuses the tyrant to throw his companions into mortal fear by the slightest suggestion, the smaller the occasion the more amusing the fright. And in general the slighter the real cause in relation to the effect produced, the more acute the pleasure, by reason of the supremacy thereby emphasized. It is always more amusing to scare a child by a slight movement of a finger than by a vigorous act of the whole body. It seems to me that it is by this association that disproportion, incongruity, irrelevance, however induced, become in themselves amusing, ludicrous, laughable. So the incongruous, in which I have no part whatever, becomes a comic spectacle and the basis of all comedy, yet also of the tragic and tragedy. In the tragic the discordance between what is and what ought to be, instead of pleasing, pains. What is comic to a coarse mind may seem tragic to the refined. A bird, distressed by the death of its mate, offering it food, might amuse a savage or a boy, but must be a pathetic sight to a civilized and cultured man, though both might be amused to see a child presenting food to its doll.

Not only may the incongruous which is comparatively unrelated to me be amusing as well as that which I myself bring about, but even when I am the victim I may be highly delighted by the intrinsic disproportion of my experience to the exciting cause. With some persons, perhaps rather few in number, the next best thing to playing a joke is having one played on them. This amusement at oneself occurs even among savages. When Stanley was on the Congo, he was at one time greatly annoyed by the number of native visitors. In vain he 367tried to repel them, but one morning when a crowd had assembled at the river side, at some little distance, waiting an opportunity to board his vessel, some of his men put on lion skins which were at hand, and acted the part so well that the intending visitors fled in abject terror. Having retired to a safe distance they looked back to see the men walking the deck with the lion skins in their hands and laughing most heartily. Seeing then the groundlessness of their alarm the whole crowd burst into roars of laughter and shouted in merriment for a long time. Exhibitions of fright, we may remark, seem to be especially amusing to savages, as when an assembly of Africans of the lowest type went into ecstasies of uproarious delight on seeing a stereopticon picture of a frightened negro hastily climbing a tree to get out of the way of the gaping jaws of a crocodile.

Play is then very largely either a mutual shamming of expression, or a stimulating real expression in one by pretended expression in the other. The pleasure in deceiving others by simulating expression points to ancestral experience, for deceit has been one of the greatest factors in life evolution. That an individual seems to be an entirely different being from what he really is, has often been most advantageous in the struggle for existence, and hence a large variety of simulated expression has been employed. Children then, as repeating in play form the race history, take great delight in masquerading and so deceiving their acquaintances as to their identity, making false pleas for charity, etc. The drama has its roots in this form of play. To make others take us for quite different than we are gives us a high pleasure of power, and early man was often moved, in the breathing spells of the struggle for existence, to play at false personalities simply for the pleasure in itself of being a successful actor. There is also the counter pleasure of the spectators in piercing the simulated expression. It is only 368in latest phases of dramatic art that simulation comes to be appreciated for its own sake, that there is on both sides full and complete feeling of the illusory nature of the whole transaction, and an enjoyment of the art *per se*. Simulating expression is the actor's art; but when the simulation is forgotten by either actor or audience, nature appears and art disappears.

While it is the province of the actor's art to simulate expression, it is in general the office of fine art to imitate and render the expressive by image, picture, musical notes, etc. The artist is the expresser and simulator *par excellence*, and complete and perfect expressiveness is his constant aim, though not for utility or amusement, but for the sake of awakening the æsthetic emotion. I cannot then agree with Bosanquet, who, as I understand, makes æsthetic feeling the emotion of expression, expression for expression's sake. For expression by its very nature is such only as *expressive*, that is, as going beyond itself, as being a means, and not an end in itself, hence expression for expression's sake is meaningless phrase. Expression, so far as it attempts to stand merely for itself, is an empty mannerism and a barren technique. Expression is only such as it is backed by the emotion expressed, as significant of some psychosis; and artistic expression or art is the expression of the artistic or æsthetic emotion, a peculiar feeling about things, as apple blossoms, a sunset, a child playing. This emotion is often awakened by cognizance of an expression, as an expression of joy or horror by a child, and may thus be an emotion of or at expression, as also in the case where it is roused by the skill in purposive expression of any kind, æsthetic or other, but expression is obviously not the only way of exciting the emotion, its object may be a mere patch of colour, a pure musical note, etc. Æsthetic emotion tends to manifest or express itself just like all emotions, and in attaining perfect expression it strengthens itself. Likewise 369language, as an instrument of thought, a logical expression, has strengthened thought, but a purely formal logic is as barren and void as a purely formal æsthetic. Language as expressive of thought and as expressive of æsthetic emotion is equally dependent upon what it expresses, and æsthetics is thus not peculiar in its relation to expression.

The interpretation of expression in nature and art is often a hard matter and has given rise to much variance. For instance, take the much discussed Laokoön group; Winckelmann says the father sheds pity from his eyes like mist upon his sons; Lessing finds grief and noble endurance expressed; Goethe thinks the father shows pity for his youngest son, apprehension for the older son, and terror for himself; Lübke finds only mere pain manifested. Coming to a single feature, the mouth, we find the most diverse interpretation. Winckelmann says that here is an heroic soul who disdains to shriek, and gives forth only "an anxious and suppressed sigh." Lessing maintains also that here is a shriek softened into a sigh, but not "because a shriek would have betrayed an ignoble soul, but because it would have produced a hideous contortion of the countenance." Later critics have generally followed Lessing. It is obvious, I think, that the expression of the mouth is not shrieking, but is moaning, groaning, or sighing. On this quite a number of competent witnesses, physicians and psychologists whom I consulted, are practically agreed. However, it has occurred to me that the sighing or moaning of Laokoön may not be a softened form at all, but the actual expression designed by the artists. It is generally supposed that the artist here desired to show mortal agony, and it is assumed that shrieking is the expression of mortal agony. This assumption seems to me correct when extreme pain is suddenly inflicted; but when, as in the case of Laokoön, the mortal wound is received only after the most exhausting 370struggle, the natural expression is moaning. The realistic sculptor would surely not give any softened form, but, shrinking from nothing, has expressed Laokoön in this death grasp in the very act of giving up the ghost. Though the muscles of the limbs and trunk are still tense, yet the closing eyes, the head falling back, and the arm thrown toward the base of the brain indicate that the struggle is over, and the death moment has come, expressed vocally only by a moan. We do not need to find here then any conflict between realism and the artistic sense, but the simplest and most obvious interpretation is what the expression gives, sighing and moaning, which is the true one under the circumstances, and is so meant by the artist.

CHAPTER XXI

CONCLUSION

In the present haste to construct psychology as a natural science cognate to chemistry, physics, and biology, we note much that is premature and confusing, owing to insufficient reflection upon the quality of the phenomena. A consciousness is a natural phenomenon, but we cannot discover and investigate it as we do phenomena of light and electricity. Anger is a phenomenon occurring millions of times every day, but it is a fact which must be discerned and studied by an altogether different method from facts of crystallization, erosion, or plant growth. Psychology is not a science of inspection, but of introspection. If I know I am angry, I know it by a direct self-awareness; if I see a man strike another, and regard this as expressive of a psychosis, and that of a certain kind, anger, this supposed knowledge is analogical realization. One who never was or could be angry could no more investigate anger than a blind man light, and, other things being equal, the more irascible a man is, the better observer of anger he would be. We are not, however, conscious of all our mental processes, and we may be often blinded to the real nature of such we think we have; and as to the psychoses of other beings, especially of the more unlike and remote, we need to be extremely cautious in forming conclusions. It is likely that the mental constitution of organisms differ as widely as the physical, that the mentality of a fish is as diverse from our own as its 372physical structure is unlike our own. The fish may have peculiar psychoses of which we may never gain the least inkling, because we cannot examine its consciousness objectively as we do its fins, its air bladder, and its gills. The psychologist must then be myriad-minded; his fitness is the ductility and range of his psychic capacity. The richness and receptivity of his own mental life must be infinite if he is to come to full knowledge of the whole course of psychism. Thus psychology is marked off from all other science as distinct in subject and method. Its being so individual and subjective is the greatest hindrance to its progress, for science is verifiable knowledge, but how shall we have a method of consciousness verification? A man tells me he has a scar on his left knee, and this I can verify by personal examination if I like, but if he says he is angry, I have no such means of verification, I can only guess by expression. A biologist announces the discovery of a pineal eye in a certain embryo, and straightway the fact may be verified by a host of observers; but if a psychologist announces that he has discovered a new mode of consciousness, the verification is by no means so easy. May not the consciousness be entirely peculiar to him? The psychologist who attempts to verify cannot disclaim the fact simply because he cannot find this act of mind in himself. But an introspective *consensus*, though extremely difficult as compared with the objective *consensus* required by objective science, is not impossible, but it requires exceptional gifts and training in introspection. Before psychology can reach any standing a method of subjective verification must be formulated and adhered to as rigidly as corresponding verification is required by objective science. The backwardness of psychology is in this most significant, that while a half-dozen recognised biologists may announce a certain fact, and it is immediately accepted as scientific knowledge, no such action can occur in psychology. The uncertainty of subjective verification is the trouble, and the 373most important step that can be made to-day is a clearly defined basis for an exact verification. That one party should claim there is a feeling of relation, and another that there is no such feeling, marks a crudeness in the most general matters, and points to psychology being about where physiology was when the circulation of the blood was in debate.

But, say the experimental psychologists, subjective verification is impossible; psychology, if it is to become a science, must, like the other sciences, resort to the laboratory, and by definite and exact experiments produce the facts to order, study them by the most approved instruments, and obtain with certainty a knowledge of their laws. Now it is sufficiently easy to experiment on light, sound, and on plant growth in a laboratory, but how can we make consciousness to order with the same certainty? how can we know when we have got a consciousness, what kind it is, etc., except by subjective verification? You certainly cannot see the consciousness or touch it; but you must wholly rely on the subjective report of the individual experimented on as verified by your own consciousness. We have no impassive agent entirely under our control, except in hypnosis, and we cannot secure conditions with the same exactness in testing the intensity of some form of consciousness, as anger, as in testing the tensile strength of iron.

In the physical laboratory we produce certain conditions and we get invariably certain observable and measurable results, but in a psychological laboratory how shall I get with certainty a definite consciousness in a large number of cases and formulate its law? How shall I know at a given moment that the mental act of the agent is what my experiment requires? Moreover, does not experimental psychology by beginning with human consciousness enter rashly upon a very complex field? If it would get results, let it start with the simpler mental life, just as biology has 374founded itself in a study of simplest elements. But how shall psychology get at the consciousness of a clam with the same exactness as biology investigates the circulation of blood in the clam? It is plain, in short, that if we are to have a fruitful experimental psychology, some very important questions of method must first be settled. A method of getting psychoses to order, to obtain the exact reaction required, and knowing and realizing what it is when got, this is a *desideratum* not yet attained. Further, we must remark that experimentation is itself a psychic act, and sense of experimentation is a disturbing factor in results; that is, a consciousness which is conscious of being experimented on is thereby complicated over mere observation method. This is markedly the case in self-experimentation. Consciousness is not, like an electric current or a sound wave, an objective fact, readily reproducible in the laboratory. And again ethics may interfere with psychical experiment. How far have we a right to incite psychosis for experiment's sake? How far may psychical vivisection be carried in the name of science? A scientist who should for his own study make an animal or person angry, would be reprobated as would the artist who should incite anger in his model in order to catch artistic effect. However, that there is a vast scope for experimental psychology cannot be denied, and we may expect an indefinite multiplication of artificial psychoses and combinations comparable to the artificial syntheses and new compounds of the chemical laboratory. Mind may develop and act merely on the scientific motive, and accomplish by *tour de force* a complex field of artificial consciousness quite distinct in origin and nature from natural consciousness. But for the present, at least, we regard not experiment but observation as the main method. Not laboratory, but field work, is most needed. The psychical scientist must go psychologizing, as the botanist goes botanizing. But there is no simple objective 375method as in botany. In order to have insight and interpretative power, there must be constant self-observation. He can know the real nature, conditions, and laws of other minds only so far as he realizes them in himself. If he has never feared, he will never know fear, and if he has never analyzed his own fear, he will not know its factors as occurring in others. All external consciousness is but a projection from the observer's own consciousness.

But it may be said that mind is but a kind of neural function, and that physiological psychology will give us the true key to consciousness. But if one has never known any psychosis, as fear, directly in himself and indirectly in others, how will he find it in any nerve activities? Nervous activities are significant of psychosis only so far as psychosis is already known. In fact, the sciences of neurosis and psychosis are radically distinct. I stick a pin in my finger, the facts of pain, volition, anger, etc., are of one order knowable only by introspection, the nerve excitation, current and reaction are of another order, constitute a complete circle, and are known only by inspection. Neurology in its own field can afford to ignore psychosis, for it does not find it: it finds only neural changes, and psychology likewise can afford to ignore physiology. These sciences stand self-sufficient, and may develop indefinitely each in its own way without meeting. Divide and conquer. The present mingling of the two is greatly to be deplored. Thus in current books we often find such sentences as this: "The prevalent view hitherto has probably been that the same nervous apparatus which on moderate excitement produces sensations of pressure or temperature, produces feelings of pain when irritated with increased intensity." (Ladd, *Outlines Physiological Psychology*, p. 387.)

This confusing of objective and subjective terms, sensation and irritation, is but too frequent in recent treatises. There is no way yet found of discovering psychic facts in 376neural, or neural in psychic, whatever may be their connection and interdependence. If we must have a cross-interpretation, the psychologist has the vantage-ground on the basis of evolution by struggle. *Nisus* has developed all sense and motor organs and all nervous organs. It is the effort at seeing that has produced the optic nerve and the physiological function of sight. The vision and visual organ of the eagle came by incessant looking for prey during thousands of years. Hence mind is not reflex or concomitant of nerve, but nerve is outgrowth of mind in the struggle of existence, and a psychological physiology is better than a physiological psychology.

The psychological field is then first, self; second, other selves or individuals. In this latter phase of human psychology we have the psychology of adults, then adolescent, senile, infantile, sexual, and racial psychology. In sub-human or comparative psychology we include animals, wild and tame, also all discussion on plant psychism, mind stuff (*e.g.* Clifford's), etc. In superhuman psychology we include all doctrine of cosmic intelligence, teleology (*vide Mind*, x. 420).

We have limited ourselves to evolutionary psychology and that of the feelings, and our data are mostly from adult human consciousness. Evolutionary psychology bases itself on the idea that mental development originates and is continued through struggle or will effort. Such evidence as we can gather points to feeling, impelled exertion as the essence of psychic evolution, and it proves fruitful when assumed as a guiding principle. And the principle of struggle is final. We cannot admit with Bain a principle of spontaneity. The activities of a new-born lamb are seemingly spontaneous only because they are the results of energies stored in ages of psychic effort. This doctrine of struggle does away with all impressionism and all passivity theories. Mind is not a receptivity, an association of impressions, a reflex or concomitant of 377physiological activities, but it is dynamic determining vital fact, an active response to the conditions of self-existence. This impetus of struggle and striving seems to feed all life and make life, and has its place, perhaps the highest in the dynamic whole we term the universe. While the significance of struggle is a question for philosophy, yet, as matter of fact, it is the only method of realization we know; and the office of humanity is the providing a wider and higher scope for struggle, the making new and independent life regions. Science and art, ethics and religion, which are at bottom only phases of emotionalism, are with utmost toil developed for themselves, and new emotions now arising and yet to arise will be cherished for their own sakes. Mind begins and continues long as the servant of the body, it ends by making the body its servant, the instrument of the spiritual life, the temple of the Holy Ghost; but all its evolution is through supreme effort. In the spiritual evolution he who loveth his life shall lose it, he whose struggle is in the primitive stage, namely, for material existence, loses thereby the real life, the life of the spirit.

It is possible, indeed, that we may over-estimate this salient fact of struggle, and certainly, in the present state of psychology, modesty is most commendable. We would be far from assuming that the horizon of our mind is the limit of the universe. However, assuming mind as a biological function continually evolving in the service of self-conservation and self-furtherance, our endeavour has been to point out the general trend of the evolution of feeling, and to analyze some of its more important features. The little exploration we have made suggests the greatness of the unexplored field of mind, the vast number of psychoses unknown, and perhaps unknowable. The difficulties of the subjective method make it seem almost impossible to trace a complete history of mind. For mind to return over and realize its whole growth in all its ramifications 378seems quite as hard as to develop new forms, or a whole region of artificial psychosis. In the filling up of missing links, psychology presents vastly greater difficulties than biology because of its subjectivity of method and the evanescent nature of the facts. Further, the more I analyze consciousness, the more I am convinced of the great and often unexpected complexity of apparently simple forms, and I am satisfied then the simplicity and completeness of the system-making psychologists, physiological or idealistic, is factitious and delusive. An inductive science of mind is yet in its infancy. My conclusion that mind was at first, and is always as progressive, feeling-impelled will, and that sensing arose as secondary, as useful cognitive effort, is simply the best reading I can make from present data when assuming the current doctrine of evolution.

A very important point, which needs to be worked out more fully than we have been able to do, is as to the nature of revival as involving emotion. Sense of re-experience and of the experienceable is one of the most important acquisitions of mind. The self-consolidation and organization of experience certainly does not come in the first place by any mechanical association, but we must assume that all mental progress is the result of the most intense, though often blind and fortuitous striving. But just how the return of an experience is cognized as *return* and as *experience*, and so becoming basis for emotion, this is a most difficult inquiry on which we have made but a few remarks in the chapter on the nature of emotion. Just when and how sense of experience is generated, and what is a full analysis of its nature, must be postponed to some future study, but I am convinced that a very fruitful field for investigation lies in this direction. Experience certainly does at a very early stage become compound, become self-appreciative in some form, as sense of the potentiality of things, but the elucidation of progress in 379this line is confronted by many difficulties. The history of ideation or representation as a power for self-conservation has yet to be traced with definiteness and completeness.

Another point, which needs a far fuller discussion than we can now give, is as to the nature of organic interaction in consciousness, as to the real quality of psychic cause and effect. We have all along assumed feeling as stimulant of will, both the will to know and the will to act, but just how does feeling develop will as struggling effort? What is the exact mode of connection? We conceive readily of physical impact as determining effects in the material world, and we conceive a transference and transmutation of energy, but in the psychic realm we have no entities as permanent existences susceptible of entering into relation with other entities. How then does a pain incite a will activity? A peculiar form of consciousness we term will activity does directly follow upon feeling pain, and, within limits, the greater the pain, the greater the willing, but we have no theory to express the mode of connection of these consciousnesses. All that we can say is that one does follow upon the other as somehow caused by it. Yet it is certain that the limitation of conscious capacity must in every individual determine a definite range of interaction, and, beyond some particular point, the more I feel, the less I will, and *vice versâ*. But the phenomenon of interference is likewise as obscure as that of excitation. The development of distinct organic forms of consciousness is slowly carried forward, and they exercise a definite dynamic relation to each other, though the mode is as yet wholly obscure. Thus the largest subdivisions of consciousness, knowing, feeling, and willing, become determined as distinct organically related modes, like the nervous, nutritive-circulatory and motor systems forming one organic whole body. These psychic modes attain gradually an intricate and definite development, whose 380constant interdependent connection with an individual body we term a "mind." And we must remark that this vital relation of one consciousness and one form of consciousness to another is in no wise effected through apperception, through a third distinct consciousness, a cognitive one, which unites them in idea. A feeling excited a will act long before there was consciousness of either, or of their relation. In general we must say that consciousness does not consciously forge for itself its own relations, but that in by far the larger part of psychic development new modes of consciousness and their inter-relations come in a totally unforeseen way, by a blind striving in the struggle for existence. It may be doubted, indeed, if even the most advanced human mind can really invent a new consciousness or a new relation in consciousness, but by intense effort it attains them. One of the obscurest points in biology is as to the nature and cause of morphological variation, and the subject of mental variation is for psychological science far more obscure. We presuppose that mental variations somehow arise in response to sudden and great emergencies, and in connection with the severest effort. Mental progress is, in all the earlier life at least, only achieved under pressure of intense pain actually experienced or ideally so,— emotion—and in some way an appropriate and saving psychosis as response of organism to environment originates. This new form may be indistinct, and proceed as a gradual differentiation from previous types, still the method of action of the motive force seems mysterious. We can see, indeed, the advantage which accrues, for example, to the animal which is first able to detect danger or nutriment by scent, but just the method of the rise and progress of scenting as a conscious process seems difficult to trace. We cannot say that power of smell arose because organs of smell were developed; this puts the cart before the horse. It is the struggle to sense that is the prime motive force in developing the sense 381organs and not *vice versâ*. We do not smell because we have noses, but we have noses because we smell. That the sense of smell is a differentiated general sensation is likely enough, but we are unable to follow the steps. We know that the higher development of our present senses is attained only through great exertion, which determines a physical basis and organic progress—as in microscopy, telescopy, and so-called mind-reading—and if humanity is to develop in the future an electric sense or a telepathic sense, it must be reached by the intense struggle of a very few. We must believe that every mode of mind is at bottom but some modification of pre-existing forms, and it may be that as all modes of the material are interpretable in motion, so the manifold mental may be equally resolvable into some one type. Yet so far as we can now see, feeling, will, and cognition seem radically and primitively distinct. The missing links in mental evolution are most difficult to determine, for, as we have often remarked, while we can with comparative ease both determine fossil organic forms *a priori* and discover as realities, the intermediate mental forms can only be known through a subjective realization.

It does not help us to ascribe the advantageous variation to chance, a word, indeed, which does not belong to the dictionary of science, for it is but a cover to ignorance. Chance means that the determinate line of causes is hidden from the observer, who only knows that one of several results will take place. Chance is thus wholly relative; the gambling of savages is often calculable to the European, and so every affair of chance, as dice throwing, might be calculable to a superior intelligence who could compute or watch every turn of the dice. Chance, then, does not reside in the outward thing, is not a property of phenomena, but is wholly a subjective limitation of the investigating mind, hence to ascribe variation, physical or psychical, to chance is simply to objectivise our own imperfect cognition. The pre-supposition of all science is that every event or 382change has its definite determining antecedents, and that these are cognizable; hence the doctrine of chance has no place in any complete and real science of phenomena. That organism is, indeed, fortunate, which first achieves some notable and valuable psychic mode, but this good fortune does not in any wise come by chance, or by the passive enjoyment of concurrent favourable circumstances, but it is a well-earned superiority attained only by the severest and most patient responsive struggle, and there in every case a determinate series of steps in mental process which may ultimately be traceable.

Mental forms also arise through perversion, competitors perverting originally advantageous variations, as has been already pointed out for paralysing-fear, sense-destroying anger, etc. Atavistic tendency gives pseudo-variations. Certain mental forms may be negative in origin, that is, merely reactionary from previous states. Given a high degree of any joyous emotion, say hope, and suddenly remove its conditions, and the swing is back beyond the zero point of emotion to actual negative emotion, as despair. Still the whole gamut from positive to negative, as from highest hope to deepest despair, is but a single generic emotion form of polar correlate elements, which have equally developed through struggle.

The subject of psychic intensity in general, and feeling intensity in particular, is likewise obscure and difficult. Physical intensity is comparatively easy to investigate in its nature and laws. For instance, in the case of light we clearly conceive its nature in terms of molecular motion, we measure it exactly by photometers, and we know it to proceed by the law of inverse squares. We have no similar certainty and clearness with regard to mental intensity. We speak of suffering very slight or very intense pains, but there is no scientific theory or valuation of psychic intensity. Mere physical intensity does not explain psychic, and we know that variations in rapidity 383of ether waves, for example, give, not quantitative, but qualitative psychic variations. 640 billion vibrations are felt subjectively as the comparatively feeble colour blue, while 450 billion gives the striking and intense colour, red. It is only within a certain range and with certain forms of forces that Weber's law of geometric and arithmetic increase applies.

Strictly speaking, we cannot apply quantitative conceptions to consciousness, inasmuch as mind has no spatiality which is the basis of idea of quantity and size. Hence the use of quantitative terms, like great, large, small, little, etc., is an indirect reference to intensity. I was in very great pain equals I was in very intense pain. No consciousness is literally either larger or smaller than another, because consciousnesses cannot, by reason of their non-spatial nature, enter into quantitative relations. So-called massive pains are really manifold. (See on this and kindred points my remarks in *Nature*, vol. 40, p. 642.)

A popular test of mental intensity, and one which has a relative value, is by the power needed to displace a given psychosis. Thus, if a man in a brown study walks into a pond of cold water without noticing it, we rightly conclude that he is thinking very intensely. This, of course establishes a scale relative to the individual, beginning with a psychosis which resists all displacing agencies, and ending with those of such very slight intensity that they give way to any and all diversions. A consciousness which supplants another must *per se* be more intense than the other. All that which rouses and diverts patients suffering from monomania and fixed ideas is practically equal in intensity. While we may thus pronounce one state as being equal in intensity to another or as being more or less intense than it, we yet have no ground for any numerical estimate. When a person says, "I feel twice as bad as I did yesterday, or I feel a hundred 384times as happy now as I was a year ago," it is plainly a general and indefinite expression. Emotions have not yet been brought within the range of mathematical comparisons.

The intensity of feelings, as also of sensations, sustains undoubtedly certain mathematical relations to intensity of objective stimulus, but owing to their complex nature, emotions, at least, must always be very difficult of interpretation by any such law as Weber's, though simple pain may be brought more easily under some law. A pain, *other things being equal*, increases in some ratio to increment of physical stimulus. But we must believe that the reason for the diversity between proportion of actual increments of stimulus and actual increments of sensation and feeling is largely physiological. It certainly is not a true psycho-physical law, a law of relation of mind and matter, as is often claimed; for we cannot obtain an absolutely objective standard to test subjectivity. Hence any such law is merely a law of relation of different kinds of sensations, of different methods of interpreting the objective. Intensity of stimulus itself is always determinable only through some sensation, which is itself subject to Weber's law. There is no objective standard for sense stimuli; the measure of increasing stimulus to increasing sensation must be by some sense which has its own law with reference to physical increment as interpreted by another sense equally under law, and so on. Take pressure, for instance; we note by sense of *sight* the arm of a balance reacting regularly and constantly to definite small additions to load, while upon our own arm we do not notice the same additions in any such series of feeling of pressure increments. The arm and balance as disparate weighers must, of course, be in certain ratios related, and for a certain range we must have a geometrical series, but other ratios at other points.

That the degree of sensitivity is proportioned to the 385intensity of sensation already present, that the knock at the door must be the louder the more noise is going on within, is a defect in organic measurement, but it is not entirely absent in mechanical; scales which weigh by the ton do not respond easily or at all to minute weights. But, abstractly speaking, mechanic methods are in general far superior to organic; a fine balance weighs better than any arm, and a good camera pictures better than the best eye; that is, their ratio of discriminating sensibility is far greater than natural organs, and it may be as geometric series to arithmetic series. Practically, however, organic weighing and seeing are well adjusted to the demands of life. An appreciation of gravity, so far as it is of use to the organism, is secured, and if a finer sensibility were demanded it would be attained. That is, I am inclined to believe that the Weber-Fechner law of definite mathematical proportions is purely empirical, and does not mark a real limit or a fundamental psycho-physical law. If a man's life and living depended on it, he could become a good weighing machine, and in time a race of organic weighers might be raised up which should vie in accuracy and range with the best scales now constructed. The quotient of sensitiveness is really indefinitely variable. It is probable, indeed, that deep sea organisms have a discriminative sensibility for both gravity and light far more delicate than the acutest human sense.

The whole subject of measurement of mental intensities must evidently be approached with the greatest care, and the diversities of researches in results and in their interpretation, is evidence that we have not completely isolated the facts we are in search of. Conscious experimentation must be allowed as tending to disturb sense. When attention is strained to marking sense increments it may very easily be deluded, and wrongly suppose as to feeling or not feeling. Consciousness is by no means infallible as to its own acts, and especially when artificial. Feelings 386may, and often do, originate subjectively by suggestion, and hence may have no direct reference to the external cause which is under experimental manipulation.

And not only have we thus to guard against a strong tendency to introspective and apperceptive error as to what we actually experience, or how we experience, but we have also to constantly bear in mind that every experience, every sensing, as of pressure, light, etc., is not an isolated phenomenon, but as resting upon and involving the past, it can never be a simple direct measure of the objective present, as a given weight or light. Every conscious experience, like all other vital organic phenomena, has thus an individuality and differs from every other as every leaf differs from every other, and so the laws of experience are capable only of general expression. Since all consciousness is self-integrating and brings up the past into itself, it is always more than any occasional reflection of a present phenomenon; in the finest analysis every consciousness must have an equation of its own.

However, there is a quotient of relation of physical stimulus, mechanically measured, with increase and decrease of both sense and of pleasure-pain. The pack-carrier feels in a certain proportion to his present load pressure of weight-increments, and pressure pains also augment, though probably not in strict corresponding ratio. It is a popular saying that the last straw breaks the camel's back, and it is certain that pains rapidly culminate. It is probable that increments which may not be sensed may yet be felt as pain. In fact, it is but very gradually that sense of pressure is evolved as practically free of pain; as a mere cognitive process it is always secondary to pleasure-pain states which are felt directly from weights or but slightly objectified. Pleasure-pain which proceeds from weights gradually is driven to sensing them—the evolution of the pressure sense—and to noting variations, sense increments, and if, like marine organisms, we ranged 387through pressure zones, the significance of discriminative sensibility might be very great.

However, it is obvious that in its rise and in its whole evolution, pleasure-pain is bound up with the pressure sense, but not with the arm of the balance as a record. Hence it is possible that Weber's law, so far as applicable, is in some measure a result of feeling interference. The simplicity of direct reaction is being destroyed by the hedonalgic law disturbing the direct ratio; we may thus feel an increasing pain from increasing weights, and have decreasing pressure sense. Beyond a certain point the law of increments, with reference to external standard for sensing and for pleasure and pain are in inverse ratio. On a very hot day we notice more and more strongly each additional degree of heat by the temperature sense, but beyond a certain degree, peculiar to the individual at the time, sense of heat will rapidly diminish as heat increases, and with increase of pain.

As to the number of feelings, of qualitatively distinct states, we must on a general doctrine of evolution pronounce this to be innumerable and indefinite. The present forms of feeling in human consciousness of course represent but a small fraction of the total number which have arisen in the course of psychic evolution. Every distinct form implies a long evolution of intermediate types which are now for the most part beyond our realization and so beyond cognition. The process of naming affords some slight clue to the importance and multiformity of feeling, though this denotes only a few of the most obvious points which have impressed themselves on the popular mind. Certainly the most striking fact to ordinary introspection, human and sub-human, is feeling, and the manifold variety of simple pleasure-pains and of emotions has always, and will always, attract most strongly the general attention. It would be a most interesting and profitable study to follow the course of language in its denotation 388of feeling. Varied expression for varied feelings is gradually achieved in vocal forms, which expressions become a language sense to denote the feeling expressed. Thus the hoarse bellow of rage will both express and denote rage. The vocal expression form as imitated is the earliest language form, and only very gradually does language assume the mechanical and arbitrary forms of its highest development. It is by imitating being mad vocally and otherwise, and pointing to the angered one, that the savage conveys the idea of anger. Gradually all but the vocal expression is dropped, and this conventionalized, becomes the origin of the word to denote the emotion in question. Feeling and emotion names are doubtless in their origin debased vocal expression forms, though in the later evolution of language this is generally not detectable, and various other more indirect associations control language. Only states of consciousness which have attained a considerable force and prominence receive notice in the vocabulary of common speech. For many variances of feeling there is no word denotation, but it may be given by intonation. The number of names of feeling is thus in any language, or in all languages, but a very rough index to the actual number of kinds of feeling, and we may expect that a thorough scientific analysis will develop as extended scientific nomenclature of feeling, as chemistry has of kinds of matter. At the present crude stage of psychology we must affirm that the number of cognizable, but unnamed feelings, far exceeds the number of the named, and that the number of the undiscriminated or the undiscovered feelings far exceeds the number of both forms.

On the whole, it has been the object of our present studies to point out with some definiteness the extent and mode of the early differentiation of feeling. Owing to the peculiar difficulties which beset this form of study and to which we have often adverted, our conclusions may seem 389rather meagre and uncertain, but it is sufficient if they emphasize a region of introspective study, which, though of the utmost practical importance, is yet the most neglected of all in psychic science; and we hope to have set forth the most probable general order of mental evolution with some distinctness as based on the struggle of existence. Mind, beginning in pure pain, and culminating on the feeling side in the higher emotions, contains an intermediate, continuous, indefinite number of forms, determined by the demands of life and preserved by natural selection, many of which are so entirely outgrown that they may be for ever beyond human conception, and many occurring only occasionally in human consciousness as survivals, and a large, yet comparatively small number constituting the present evolution phase of feeling in human consciousness. We have dwelt specially on the lower developments, the rise of objectification and its nature, the rise and value of emotion, with some characterization of the simpler and earlier emotions. Emotion is superior to and supplants sensation, though based thereon. The poison I fear, I abstain from without tasting; but with lower psychisms there must be a direct sensing of the thing before its experience quality is apprehended.

Must we not suppose that feeling and emotion is destined to be an evanescent form in the evolution of mind? Is not the emotional type gradually disappearing, and will not the men of the future be pure indifferentists? Or are we rather to judge that emotion will always continue to strengthen and deepen? In an intellectual and introspective age like our own the naïve mental life, which is primitive and merely natural, vanishes, and we find that men everywhere, like Kenyon, in Howell's novel, *The Undiscovered Country* are constantly destroying their feelings by pulling them up by the roots to see what they are and why they are. Such are only occasionally surprised into a genuine emotion, but they greet it with joy, 390and forthwith pull it to pieces in a morbid self-analysis. An indifferentism, born of intellectual curiosity, of scepticism or of pessimism, is, in fact, a pathological state, a certain mono-emotionalism, for humanity cannot escape emotionalism if it *would*. This *blasé* way of looking at things and *feeling* about them, is a decadent symptom in an artificial age. The struggle of life in a natural state always demands a varied, prompt, and frank emotionalism. If mind lose its love of things and men, it may yet be moved to highest attainment by sentiments like the love of science and truth. An intense intellectual life must be driven to its strugglings and achievements by some strong motive power, some powerful emotion, though this may be purely impersonal, like the conviction of duty, or the love of truth. Feeling as the fundamental element in mind, as the very core of mentality, as the force which actuates both will and cognition, can never be destroyed, and for the future progress of mind, as for the past, we are assured that feeling and emotion will not cease to become ever stronger, deeper, and nobler.